Rubik's Cubic Compendium

Recreations in Mathematics

Series Editor
David Singmaster

Rubik's Cubic Compendium

Ernö Rubik, Tamás Varga,
Gerzson Kéri, György Marx, and
Tamás Vekerdy
Translation edited by
David Singmaster

Oxford New York Tokyo
OXFORD UNIVERSITY PRESS
1987

793,74
R824

Oxford University Press, Walton Street, Oxford OX2 6DP
Oxford New York Toronto
Delhi Bombay Calcutta Madras Karachi
Petaling Jaya Singapore Hong Kong Tokyo
Nairobi Dar es Salaam Cape Town
Melbourne Auckland
and associated companies in
Beirut Berlin Ibadan Nicosia   89 - 0517

Oxford is a trade mark of Oxford University Press

Published in the United States
by Oxford University Press, New York

British Library Cataloguing in Publication Data
Rubik's cubic compendium.
1. Rubik's cube
I. Rubik, Ernö
793.7' 4   QA491
ISBN 0-19-853202-4

Library of Congress Cataloging in Publication Data
Büvös kocka. English.
Rubik's cubic compendium.
(Recreations in mathematics; 3)
Translation of: A büvös kocka.
Bibliography: p.
Includes index.
1. Rubik's cube. I. Rubik, Ernö. II. Title.
III. Series.
QA491.B8813 1986   793.7' 4   86-8699
ISBN 0-19-853202-4

Set by Cotswold Typesetting Ltd, Cheltenham
Printed in Hong Kong

CONTENTS

INTRODUCTION
The fascination of Rubik's cube

For Ernö, in appreciation

David Singmaster

When Ernö Rubik first thought of the Magic Cube in 1974, he knew it was a great thing. But even he could not foresee how great it would be. In June 1979 I wrote the first newpaper article outside Hungary on the cube and I predicted, half jokingly, that the cube would sweep the world like Sam Loyd's famous 15 Puzzle in 1880 or the crossword puzzle. By 1981, it had done so and much more. In many countries, nearly every schoolchild has a cube!

'Rubik's Cube' has become a household word to the extent that it has now been entered in the *Oxford English Dictionary*. It is the centre of a multimillion dollar industry; the subject of pop songs and lectures to learned societies; the cause of marriages and divorces; the cause of cubist's thumb and Rubik's wrist; a cure for tennis elbow; the subject of several best-selling books in many languages; a motif for tee-shirts, badges, and car stickers; the subject of dozens of imitations leading to scores of lawsuits; the inspiration of many further puzzles; an analogy for quarks in particle physics and the inspiration of a complete cubic cosmology; the theme of thousands of articles in journals, newspapers, magazines, radio, and television; the inspiration of several cartoons; the subject of clubs and contests and a replacement for worry beads. The cube has changed thousands of lives and made the fortunes or reputations of several companies and individuals. It has awakened hidden geniuses in the doldrums of school and made them into popular heroes and best-selling authors. It has received several 'Toy of the Year' awards and is already widely acclaimed as the Toy of the Century. It is already a permanent part of our culture.

Why is Rubik's cube so popular? This is a question which many people ask. It has no simple answer, but I will try to explain the important features of the cube and how they make the cube so fascinating.

Rubik himself has pointed out some features which he considers important.

(a) The parts of the cube stay together—in contrast to many other moving-piece puzzles.

(b) Several pieces move at once—again in contrast to all other moving-piece puzzles that I know of.

(c) The pieces have orientation. By this, we mean that a piece can return to its original position but in a distinguishable orientation from its original one. Most puzzles with this property are assembly puzzles which are very different in nature from the cube.

These are indeed important features, but they are very sophisticated ones, more comprehensible to cognoscenti than to the general public. Other features, perhaps too obvious for Rubik to mention, should be noted.

(d) The three-dimensionality of the cube is its most obvious characteristic and one which should not be overlooked, as it underlies many of the other features of the cube. It is especially noteworthy as the only real ancestor of the cube is Sam Loyd's 15 Puzzle which is two-dimensional and this two-dimensionality limits it greatly. Three-dimensional moving-piece puzzles are very rare. The only such puzzle I know of is a sliding cube puzzle of Piet Hein which is so rare that both Rubik and I recently re-invented it before learning that it had been done by Hein.

(e) The cubicality of the cube. The cube is the most basic three-dimensional shape. I have handled magic spheres and they are much less satisfying than the cube. On the sphere, it is difficult to make a 90° turn as there are no edges to align visually or tactually.

(f) The colours of the cube. This is so obvious as to be easily overlooked, but it is the basis of its great aesthetic appeal. The many cubes with other markings lose this appeal.

I think the colour patterns would be even more striking if the three primary colours were opposite their respective complementary colours. Rubik has already tried this but he found that a white face was essential to brighten the appearance of the cube. The colour pattern has now been standardized so that opposite faces differ by the colour yellow: white–yellow, red–orange, blue–green. This shows the care that Rubik has taken with all aspects of the cube. My idea of complementary opposite colours might work if the cube were made in neutral grey.

(g) The mechanism of the cube. This is the most wonderful aspect of the cube and I will say quite a bit about it. At first it seems completely impossible for the cube to work. After I first wrote about the cube, a student told me that a friend had telephoned her about a hoax article on a cube which turned in all directions. The friend announced that she had proved that such a cube could not exist!

Even when you have a cube in your hand, you cannot see how it is held together. Very few people can propose a possible mechanism and very few of these are practical. I have seen only one proposal which was virtually identical to Rubik's and this was put forward by John Gaskin, a UK patent agent, as a result of my first article on the cube.

There are several rumours of earlier cubes. A French inspector-general, M. Semah, reports having played with a cube in Istanbul in 1920 and in Marseilles about 1935. A mathematics teacher named Gustafson in Fresno, California, had the idea of a $2 \times 2 \times 2$ magic sphere in 1960, but apparently never found a mechanism for it. [He eventually did find a mechanism and patented it in 1963—DS.]

There are two patents from 1970, but one mechanism is impossible and the other is impractical. There are several patents after Rubik's, but for less practical mechanisms.

How is Rubik's mechanism more practical than the other mechanisms? The other mechanisms are all based on tongue-and-groove connections between pieces as in Figures (a), (b), or (c).

       (a)   (b)   (c)

Such connections are difficult to make with sufficient accuracy so that the faces will turn easily. Rubik's mechanism has connections as in Figure (c), which greatly simplifies the shapes of the pieces. The tongue and groove methods can be made so that the 26 exterior pieces of the cube hold together with no central piece at all. Such an attachment will not be very rigid and will become loose with a little wear. In Rubik's mechanism, it is essential to have a central piece to hold the whole structure together. The six face-centre pieces are attached to a central piece with spring-loaded screws. This gives a firm structure to the cube, while leaving some flexibility so that manufacturing tolerances need not be as precise as would otherwise be required. The tension in the springs will take up some of the slack caused by wear, but it turns out that the corner pieces wear fastest so this feature is not as useful as had been hoped.

(h) The complexity of the cube is remarkable. It is unprecedented for such a simple-looking puzzle to produce such complexity and for such a complex puzzle to be so popular. Apparent simplicity concealing great complexity is a characteristic of many real problems and of masterpieces of art, literature, etc. The cube is so complex that it is the most difficult problem that many people have ever encountered. All humans need and enjoy mental challenge. In the modern world of increased leisure and less challenging schools and jobs, many people find the cube exercises and stretches their mind in a new and enjoyable way. At first, no one thought that the cube would interest children under 12 or 14 years, but the popularity of the cube with children from 7 years shows how much we have underestimated the mental capacity of young children.

The complexity of the cube has led to several hundred solutions appearing, ranging from single sheets of formulae and/or diagrams to full-size paperbacks. Dozens of computer programs have been written to help solve the cube. The complexity of the cube is so great that everyone can find a different solution!

(i) The mathematics of the cube. This is a special aspect of the complexity of the cube, but is a consequence of some of the other features, notably (a), which make the cube into a very elegant example of a permutation group. Features (b) and (c) give the group so much complexity that one really must use mathematics to understand the cube's potential. There have been several research papers and one book on the mathematics of the cube, with more soon to appear. The group of the cube has been entered as a standard example in CAYLEY, the world's foremost group theory program. The mathematical aspects range from simple investigations suitable for school students up to research problems leading to general unsolved problems in group theory.

(j) The educational value of the cube. This is a somewhat unexpected feature of the cube, but one must remember that this was the source of Rubik's inspiration in the first place. I believe the cube has already greatly increased the three-dimensional abilities of students of all ages from 6 years old. Two years ago, such children would not have known what an ordinary cube was. Now they are happily turning them about in their hands and assimilating the basic relationships and symmetries without conscious effort. Older children are learning the concepts of group theory, both unconsciously and consciously, as they grope their way towards understanding the cube. Many combinatorial ideas and many basic techniques of mathematical thinking arise naturally and are useful to students in all scientific fields and in many other creative fields.

Having examined the significant features of Rubik's cube, we are still left with the question of why these features make such a fascinating object. Although one cannot provide a simple answer to such a complicated question, I think one can say that it is the multiplicity of different features on many different levels which makes the cube so fascinating. Every masterpiece, in any field, defies simple interpretation. Every viewer, listener or player finds his own interpretation, his own joys, his own beauties, his own favourite aspects, his own associations, his own extensions. Some find the cube fascinating for its mechanism; some for the aesthetics of its shape and its colour patterns; some for the challenge of solving it; some for its mathematical complexity and beauty; some for the kinaesthetic pleasure of turning it; some for its competitive aspects; some for the further possiblities that it opens up. Everyone finds his own fascinations.

I have already mentioned Sam Loyd's 15 Puzzle. There are two other puzzles which I would possibly mention as masterpieces: Solomon Golomb's polyominos and Piet Hein's Soma cube. Rubik's cube combines the fascinations of these past masterpieces in a way that puts it in a new category of its own—the masterpiece of masterpieces.

1

IN PLAY

Ernö Rubik

1.1 About writing books

All my life I have had reservations about making confessions, but the experience I have had now impels me to do so. It was an experience for me to find a rough diamond and to discover, while working with it, an unimaginable multitude of possibilities. I was captivated from the beginning by an interest of an intensity rarely experienced. My cubic puzzle grasped people's fancy, forcing them to play, to work, and to think. Since my Magic Cube has become popular, I have spoken to many people about it, including journalists who, naturally, want to know all about it. While listening to their questions and trying to answer them, I realized that no other person is capable of describing the motives that led me to devise the cube. If they try to do so, some sort of distortion, incomprehension, or misunderstanding is bound to occur. This is only natural, for the journalist is expressing himself when he writes an article.

I am an active person. Activity is my element. Writing is an extremely passive form of action for me. But how different and how wonderful it is to bring something into existence, to give form to a substance, to make it purposeful and attractive to the eye and the hand. So why did I, in spite of everything, decide to write about it myself? I must admit my motives were rather selfish—I wanted to understand myself. After all, the easiest way to understand something yourself is to try and explain it to somebody else. This was something I experienced during my school years. I remember an old joke about a teacher who wants to explain something to a pupil. First he introduces the problem, then he gives the solution and asks his pupil if he has understood. 'I'm afraid not,' the child answers.

1

The teacher gives a second explanation and asks again, 'Have you got it now?' The child shakes his head again. The teacher gives a third animated explanation and goes into great detail. On finishing, he asks with relief, 'Now you understand, don't you?' 'No', the child answers resignedly. 'How can that be,' says the teacher, 'for now even I understand it.'

Excuse my stale joke, but though old it is still to the point. If we want to explain something, we must speak coherently. When we are thinking on our own, we are apt to be inexact and careless and leave contradictions unsolved, but we cannot do this when we talk to others.

1.2   This is how it began ...

My story goes back to secondary school. I attended a secondary school that specialized in fine arts. Among other things, we learnt descriptive geometry. Unlike my schoolmates, I understood and enjoyed this subject. It was strange that the students of this school — among them future sculptors and industrial designers, professions that both require a complex perception of space — found the language of geometry so alien.

At that time I wanted to be a sculptor but in the end I decided to study architecture at the University of Technology. Here I continued to learn and enjoy descriptive geometry, despite the fact that the first exam I took in this subject was almost a disaster. This was my very first university exam and I felt so confident that I decided to sit the exam early. Anyway, I did not get the excellent marks I thought I would. The way in which they taught descriptive geometry did not inspire me. But my university years increased my interest in structures and constructions.

The decisive step came while I was attending the School for Industrial Design. Here I took a specialized subject, the study of forms, aimed at developing skill in construction, a feeling for plastic art and a familiarity with the hidden possibilities of materials. While making our experiments, we never asked if our products had any practical use, so we were not influenced or restrained by considerations of function. It was a great pleasure to carry out these experiments and I enjoyed the subject very much.

Later on, the Department of Interior Design at the School invited me to organize and teach the subject to graphic art students who had not studied it earlier. It was certainly high time they did, since graphic art, especially the problems of applied graphics arising in everyday life and practice, needs good spatial orientation and skill in spatial moulding. As a young lecturer, I threw myself into the work with great zest and searched for a method that would make effective use of the short time (one term) that we had.

Being at the Department of Graphics obviously meant that we worked with paper. It was clean, extraordinarily simple and easy to manipulate, thus assuring an infinite number of experiences. I am convinced that this type of activity should be done informally, for fun, and that one should not be afraid of making mistakes or being faced with possible failures. So I did not fix anything beforehand apart from the outlines of the course. I knew that I wanted to start with plane structures and then move, step by step, towards spatial structures. I constructed the individual exercises during the course myself and enjoyed the work tremendously. I am positive that the best and most realistic way of getting students involved and interested in a subject is to let them see that the teacher himself is interested in solving a particular problem and, despite his practice and knowledge, also has to work hard to reach a solution. If they see that no ready solutions, no set patterns exist, and that the teacher too must think things over before he can solve a problem, they will not feel paralysed by his superior knowledge and will take an active part in the shared work. Results and new discoveries will be achieved by common effort — by both teacher and students working together. The feeling of reward and satisfaction will also be shared.

The first exercise was the 'step-by-step manipulation' of a square piece of paper. First we bent and folded it. Then we used a knife and scissors to cut it and, later, to cut something out of it. Finally, the use of glue was also permitted. These exercises were intended to illustrate to the students my experience that although a problem is limited, it does not exclude the almost infinite abundance of possible solutions. While manipulating the material, more and more restrictions were lifted. As they worked, I insisted that my students searched for exact, clearly constructed, and easily interpreted forms.

Another type of exercise was to build structures from given elements. These exercises were intended to illustrate that by adopting different methods of duplication and construction, even the simplest elements allow us to build an unexpected abundance of forms and spatial structures. If, for example, we prepare in advance card-shaped elements of identical size and fold them, we get a series of spatial elements. Experimenting with all the possible ways of joining the cards, one arrives at a number of very interesting constructions. I believe that this kind of exercise develops the ability to understand, use, and independently describe a system of rules.

We also dealt with the cutting up of solid bodies. (These exercises were perhaps the most nearly analogous to the problem of the Magic Cube.) The most instructive way of getting to know the inherent structure of any 'solid' body is to separate it into parts following some previously conceived plan, and then to build it up again in order to check our understanding. Regular bodies are most suitable for this exercise. Their structure is clear and symmetrical, making it possible for beginners to interpret them. One interesting problem is to resolve a regular body into a given number of congruent parts. While performing these exercises, the relationship of regular forms, the possible ways of constructing regular bodies, their interrelation and some interesting but concealed properties, become clear and understandable.

Colour had an important part to play in these exercises, a factor which is also relevant to the cube. The spatial forms made of white paper were coloured but our colouring was never arbitrary. Its character had to be in accordance with that of the form. Colour must not contradict the form but interpret and enhance it. A new operation is not worth carrying out unless it adds something to what we already have.

One more aspect of my former work can be regarded as leading up to the cube. Besides teaching the study of forms, I also taught a preparatory course in descriptive geometry for students wanting to take the entrance exam for the Department of Interior Design at the art school. I supposed my students to have a good sense of space orientation (which is not at all the same thing as a practical knowledge of stereometry). As this is indispensable in interior design it was a natural assumption on my part. Yet it was astonishing and dispiriting to see how little the students had acquired at secondary school.

Descriptive geometry is a language that succinctly and clearly communicates interrelations which are difficult (or even impossible) to express in any other language. As designers, we must know this language if we want to be able to discuss our ideas about space or spatial objects before giving the ideas material form. If we want our ideas to be given form by others, we have to transmit our ideas by means of plans and instructions. We must also be able to understand other people's ideas when they are expressed in the language of descriptive geometry and be able to interpret their concepts.

When one plans an object or a space, whether as an architect or an interior designer, one must clearly perceive and appreciate the space that is to accept the plans. One must know the possibilities and limitations of the construction and assembly of spatial objects.

Related puzzles

So far, I have been talking about teaching, or about the fun aspects of science. Now I want to mention some 'real' games which interest me and which are closely related to my way of thinking. I liked and still like to play with them.

The tangram is a puzzle originating from ancient China. You cannot imagine a simpler game and yet it has a vast number of possibilities. It is a kind of jigsaw puzzle consisting of a square divided into seven smaller polygons, of different shapes and sizes: five isosceles right triangles, a square and a parallelogram. From these, a vast number of aesthetically mathematically interesting figures can be constructed.

Figure 1.1 Tangram

There is another puzzle based on the square – polyominos, invented by Solomon W. Golomb. The pentomino version contains twelve different shapes, consisting of five small squares joined together in all the possible combinations. Many interesting problems can be solved with them: for instance, one can construct rectangles consisting of 3×20, 4×15, 5×12, or 6×10 squares, or

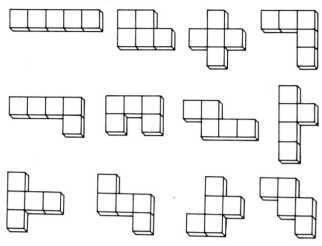

Figure 1.2   The pentominos

an 8×8 square with a 2×2 square missing at the centre. Playing with these pentominos develops combinatorial thinking.

Instead of the two-dimensional pentominos, one can use cubes as the basic elements, thus obtaining a three-dimensional version (for example, one can put together a $3 \times 4 \times 5$ rectangular prism).

Piet Hein's game, the Soma cube, is closely related to this three-dimensional version of pentominos. Here we have seven pieces, six consisting of four small cubes, and one of three, joined face to face. With these seven pieces a $3 \times 3 \times 3$ cube can be assembled. This game resembles the Magic Cube both in shape and in the large number of possibilities obtainable. There are 1 105 920 ways to assemble a cube [but there are only 240 ways if the pieces are solidly coloured—DS].

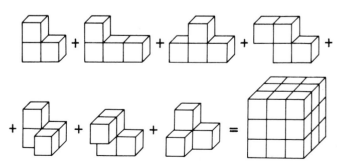

Figure 1.3   The Soma cube

Let me point out one feature of the Soma cube. The $3 \times 3 \times 3$ cube consists of 27 units (small cubes), so in addition to the six pieces consisting of four cubes, a seventh piece, made up from three cubes only, is required. This, however, makes the game somewhat inhomogeneous. In my version of the Soma cube the small cubes may be joined by their faces *or* by their edges, which means that we can have nine pieces, each consisting of three units (small cubes), since 3×9 gives exactly 27 units. I would now like you to consider this question: do you think you can build a 3×3 cube from the nine pieces? And if so, how? Is there only one possible solution or are there several?

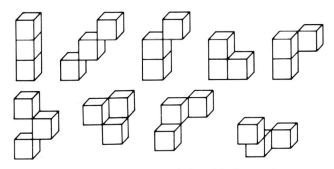

Figure 1.4 Rubik's variation of the Soma cube

MacMahon's game is also made up of cubes. It is not well known as a game but it offers an interesting mathematical problem. There are 30 cubes, whose faces have six colours, in all the possible permutations. (There are exactly 30 different possibilities.) The basic exercise is to choose one cube and then use eight others to make a $2 \times 2 \times 2$ cube that has the same arrangement of colours as the first cube, with each face a single colour [and with the interior faces matching in colour—DS]. A $3 \times 3 \times 3$ cube can also be put together, but this is more difficult.

A favourite game of my childhood was the 15 Puzzle invented by Sam Loyd about a hundred years ago. It became extremely popular in the United States and later all over the world. It is one of the most widely known puzzles. I am sure everyone is familiar with it so I shall describe it only briefly. It is a 4×4 square with 15 small squares, numbered 1, 2 ... , 15, and with one empty space. Using the empty place one has to restore the original order from any scrambled pattern by shifting the small squares.

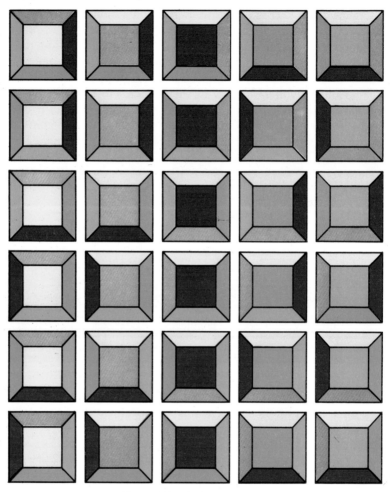

Figure 1.5 MacMahon's cubes

1	2	3	4
5	6	7	8
9	10	11	12
13	14	15	

Figure 1.6 The 15 Puzzle

Many people see similarities between the 15 Puzzle and my cube. It is true that the former was also very popular in its time and that scrambled patterns have to be restored in both puzzles. However, there are some characteristic differences as well. The 15 Puzzle is two-dimensional, and is played by moving squares on a plane; my cube, on the other hand, is a spatial game operated by turning.

1.3   The Magic Cube

After this rather long introduction, it is high time to tell the story of the birth of the Magic Cube. It began with my experimenting with geometrical elements and with the possibilities of three-dimensional constructions. Many experiments were carried out. I tried to make models of the basic forms, analysing them, resolving them into their constituent parts and then finally reconstructing them. While experimenting in this way, I got the idea that the cube, which has so many possibilities for games and puzzles, must reveal something new. It seemed remarkable that by putting $2 \times 2 \times 2$ (that is, 8) cubes to form a larger cube, its faces could be turned around three different axes (at least in principle), and the original form (a $2 \times 2 \times 2$ cube) could be restored after each 90° turn.

This, of course, seemed feasible in principle. So the next problem was to construct a cube on which all these turns could be realized in practice. It was the problem of construction that attracted me at first. I started with the most obvious way of joining them. If you join each small cube with its opposite counterpart at their corners, and fit the four couples of cubes together to make a $2 \times 2 \times 2$ cube, they will not come apart. It was very simple to make this construction. I drilled a hole diagonally through eight wooden cubes and passed strips of rubber through each of them — my first working model was ready. What I really wanted to know was what happened to the small cubes when they were turned. So I marked them by using some coloured paper which was to hand at the time, gluing the paper onto the visible sides of each small cube. Then I turned the cube several times, using all possible turns, and the colours became beautifully mixed up straight away. Now I wanted to restore the cube to its original state, so I started trying to turn it back. It was an astonishing experience for me — an experience that since then, I think, millions of people have shared — to discover that it would not work! The problem seemed to be so simple, yet it was extremely difficult. I tried to work it out, but the model fell to pieces. The rubber strips crossed one another, got tangled up and then finally broke. But the construction problem excited me so much that I decided to try and solve it. I must admit it was not an easy job. At first, it seemed to be almost impossible to find a construction where all moves and combinations of moves could be performed in every situation.

The basic question was: what would keep the small cubes together? I had to find a means of keeping them together which would allow them to leave one section of the cube and join another, rotating in a different direction, during the performance of the different turns. Many alternative solutions presented themselves. The use of magnets seemed to be the most obvious one. I made the inside face of each piece concave, in such a way that they would fit round a steel ball in the centre. Then I fitted magnets into the concave face of each piece. They clung to the steel ball, and could be moved about with the desired effect, but this was not the real solution. Magnetic force decreases as the distance of the surfaces clinging to one another is increased; the closer the two surfaces are, the larger the force, but even a small increase in distance results in a relatively great decrease in the magnetic attraction, and the cube will then fall apart. Further, this method of construction would produce a toy that would come apart and this was not the sort of real solution that I was looking for.

I wanted a set of elements that would not come apart but which would permit halves of the cube to rotate in relation to each other. If dismantling were permitted, the problem would become trivial— we could simply take the cube to pieces and then put it together again in the desired order. So I discarded the magnetic solution. The best feature of the final construction, I think, is that the rules of play are built into the construction. The player does not have to remember to keep to the rule that only turns are permitted, while dismantling and putting together are not permitted—the cube itself forces us to keep to the rule.

But what is the solution? If we are rotating a face, the rotation has an axis and although this axis does not change position, the corners do. Thus there is something that is constant, and something that changes; in other words the position of the axis is constant, while that of the moving elements changes. To solve the problem, I needed an intermediate element. So I concentrated on the edges. The geometry of the cube implies that its corners span three edges and its centre-pieces four. But even the edges change their position. The edge belongs to two different faces and if we turn one of these faces, the edge will join a third. This implies a constantly changing connection, but this is impossible unless the overall pattern of these connections remains unchanged and the role of one element can in any situation be taken over by any other element of the same form. Rotations are circular moves, so a circular track is required. Once the circle is filled up by our elements, the latter can be rotated in a full circle. What *is* important now is that no change of axis may take place until a 90° turn about an axis is made. The original shape is then restored; the position of some elements has changed but their positions are occupied by elements of the same type, and the structure as a whole does not change.

From what I said before, the track must be a surface of revolution. In the final solution of the $3 \times 3 \times 3$ cube, it is cylindrical. The $2 \times 2 \times 2$ cube has two different versions; I used a cylindrical surface in the first and a spherical surface in the other. Both of them have advantages and disadvantages. In fact, it was the construction of a $3 \times 3 \times 3$ cube that was solved first; the solution then applied to the original $2 \times 2 \times 2$ problem.

Having found the ideal geometrical form did not mean that the final construction had been obtained. It was very enlightening to see that pure geometry is not enough if one wants a working mechanism. Geometry is based on absolute precision and exactness: edges should be one-dimensional, equal sides should be perfectly equal, a right angle should be exactly 90°, etc. A physical object can only approach perfection, and if the object is intended for mass-production, it must be designed to work in spite of slight variations. It was the shape of eroded river pebbles that put me on the right lines. Here sharp edges have been rubbed and smoothed away in the course of time, and rounded shapes of extraordinary beauty have come into existence. In my case, similar rounded elements were necessary.

But our problem was still not solved. If we built the model in this way and made its joints very precisely without gaps, it still would not work, as the friction would impede its movement. On the other hand, if we built it with loosely fitting components, it would rattle unpleasantly. It would not fulfil the goal I had set myself, namely, to construct an object which, although it consists of many pieces, forms a homogeneous and closed whole.

In the final solution, I used spring-loaded screws that pulled the face-centre pieces, under constant tension, towards the centre of the cube. These pieces gripped the edge-pieces and these, in turn, gripped the corner-pieces. A sort of surface tension arose, a 'round' capillary formation similar to that of the water drop floating in a state of weightlessness. I had great hopes for this simple mechanism but even so I was astonished to see how perfectly it worked when I manipulated it. This was the moment I had been waiting for—the cube was born.

It was born—but still unclothed. It had to be clothed so as to give all important information about itself yet hide the non-essential details. The visible surfaces of all the elements (the small cubes) appeared identical. If we want to follow the amazing moves we were talking about earlier, then the small cubes have to be individually recognizable; no change would be noticed if one rotated a system of cubes which all look the same. So we had to mark them. This could have been done in many ways; colouring seemed to be the most attractive. Six colours makes each of the faces of the large cube distinguishable, and even the individual pieces become unique. The centre-piece of each face has one of the six colours, each edge-piece has two colours and each corner-piece three. The combination of colours on any piece is different from that of any other piece.

The colours of a piece tell us its correct position in the ordered arrangement. Turns of the centre-pieces on their own axes are not discernible, so the total number of variations is divided by a factor of 5000 [actually 4096—DS]. But I did not mind this in view of the extraordinarily large number of possible variations still remaining. It was not an important reduction, and the aesthetic homogeneity of the object—the simplicity of the colouring—seemed more important to me. Since that time many different suggestions have been made for colouring the cube according to other principles. None of these suggestions, although interesting, improved upon my original version. But let me describe a system of colouring that shows up the turns of the centre elements. Only four colours are used. The centre elements all have four colours, in each of the six possible arrangements. The edge elements are given two colours; there are six possible combinations using four colours, and each combination has right-hand and left-hand versions, making 12 in all—and there are exactly 12 edges. Each corner-piece uses three of the four colours; this can be done in eight different ways, since three colours have two different cyclical permutations. (This means that every corner will have a symmetrical counterpart.) Each side of

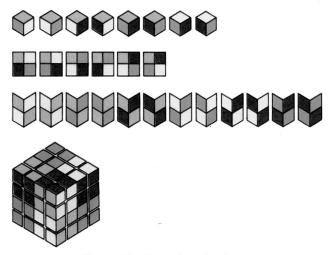

Figure 1.7 Alternative colouring

a cube coloured in this way will consist of four squares in different colours when the cube is in the unscrambled position.

The cube obtained in this way may be interesting, but the gains are outweighed by the losses. Certainly, less colour is used to give more information, but decoding becomes more wearisome, and the original visual clarity and simplicity is lost.

1.4 The family of cubes

The principle of construction I found can be extended to other spatial forms and to other systems of axes as well. I had thought this over speculatively at the very beginning. There are some interesting forms among the variants; indeed, some of them have new aspects and imply new mathematical problems. But none of them is as attractive as the cube in its simplicity. Two broad groups of possibilities can be mentioned.

The first group, including the Magic Cube, is based on a rectangular system of coordinates. The angles of rotation are right angles and so are those of the surfaces. The structure does not depend on the outer form but on the surfaces which rotate on each other. Preserving this regularity on the one hand and altering the number of axes and moving elements on the other, we can get a broad idea of the possible versions. The smallest one has $1 \times 2 \times 2 = 4$, the largest (the giant cube) has $5 \times 5 \times 5 = 125$

Figure 1.8   The family of cubes

elements. One, consisting of $2 \times 3 \times 3 = 18$ elements, has been put into production as the Magic Domino.

The $2 \times 2 \times 2$ cube, consisting of eight elements, also has some interesting properties and playing with it offers good training for solving the Magic Cube. Increasing the number of elements may be interesting, but in my opinion, it does not really add anything new.

The $3 \times 3 \times 3$ cube has three types of elements and these symbolize the three components of the cube as a geometrical body, namely its faces, edges and vertices. They are symbolized by the centres, the edge-pieces and the corner-pieces respectively. Thus, the Magic Cube gives the maximum amount of information with the minimum number of elements. If the number is decreased, some of these qualities will be lost; if it is decreased, no new quality is gained.

Another group of possibilities arises if we leave the rectangular system of coordinates. No theoretical or practical difficulty will arise if we cut up other regular bodies in accordance with their symmetry, and apply our construction principle to them. We again get a unified assembly of elements that remains together while it can be turned along the surfaces of intersection. The cube is very homogeneous. When we cut up the cube, we get cubes and only cubes as the building elements. So we are able to construct the cube using cubes only. No other regular body is so homogeneous. The tetrahedron and the octahedron can be built up from smaller tetrahedra and octahedra but the dodecahedra and icosahedra cannot be partitioned into regular bodies.

In my opinion, the octahedron, the 'double pyramid' (especially the simplest version consisting of $6 + 8 = 14$ elements) would be worth manufacturing. I think the reader will meet it in the foreseeable future.

1.5 The puzzle

Restoring the cube is an activity that can be learnt. It was not my original intention to 'teach' this type of activity to others, and I still do not have the inclination to do so. Since the advent of the cube, people all over the world have asked for help, saying that as I posed the puzzle, I must know the best solution. But this is a misunderstanding! To formulate a problem and to solve it are not the same thing. For me, it was more interesting to find the problem than to solve it, although, of course, I know that practice in solving problems helps us in formulating new ones.

I must emphasize the fact that to invent the cube is one thing, to solve the problem posed by it is another. Of course, while designing the cube I was also intrigued by the problem of unscrambling it and, with some hard work, was able to find a solution. I was forced to do so, since I had to prove to others that a solution existed, and that when I had found it I was satisfied with it. When I say solution, I mean a practical one, of course, since it is obvious at once that the problem can be solved theoretically. We always start with the ordered pattern so, theoretically when we turn the cube to get a scrambled one, we should be able to perform the same movements in the opposite direction and order, and so return to the ordered pattern. Theory and practice are not, however, identical. First, my problem was as follows: are there other ways of getting back to the original state from a scrambled one, or are we forced to use the same operations that led us to it? If the latter, it would mean that the practical solution of the problem was virtually impossible, since there are so many possible states of the cube and a wide choice of moves in each situation. If there were only one way of returning to the original state, finding it again would be hopelessly difficult. Fortunately, the nature of the problem shows that this pessimistic conjecture is wrong. Since the cube has become well-known, many methods of restoring the original pattern have been found. It is instructive and interesting for me to see that there are a lot of different methods for solving the problem, and none of them can be considered the best or the worst. The perfect solution, jokingly called 'God's Algorithm' by mathematicians, is beyond our reach, as its name implies, but mathematicians involved are working to get as close to it as possible.

1.6   The story

I cannot tell the story objectively any more than a father can speak objectively about his own child.

Objects for me are not simply objects; whether they are natural objects expressing the beauty of nature, or artefacts of human skill and intellect, in me they are always triggering memories, emotions and ideas. This is particularly so with the cube.

The life of an object is very similar to that of a living creature. It takes part in the eternal cycle of birth and death. I have already given a detailed account of the most dramatic and mysterious moment of this process, namely the birth of the cube. Here I would like to give some factual information. The idea of actually making the cube, the knowledge that it was possible and worth doing, was born in the spring of 1974. This was the most decisive step. All the subsequent events and stages of progress, namely solving the problem I set myself, producing and marketing the cube, and, finally, the cube's popularity were, in a way, a matter of course.

The cube had already been made in its final form when I realized that it was more than a tool for theoretical purposes — it had originally been conceived for illustrating spatial moves — it was also a good game that could be marketed. So on 30 January 1975, I applied for a patent at the Patent Office and started to look for a manufacturer. Luckily, I soon found one — the Politechnika Co-operative. In 1977, the cube was put on the market in Hungary. So it started out on its own life and became popular without being advertised, through its own merits. Since that time, demand for the cube has increased enormously. In 1980 almost one million were sold in Hungary alone, which is considerable in a country of only ten million inhabitants.

Distribution abroad started somewhat later. Since the first months of 1980, the Hungarian export enterprise Konsumex has exported a large quantity of cubes through an American firm, the Ideal Toy Company.

The cube won a BNV prize at the International Budapest Fair in 1978 and an award from the Hungarian Ministry for Cultural Affairs in 1979. In 1980 it won several prizes in other countries, including Germany and France. It also won the 'Toy of the Year 1980' in England; this astonished me very much since this prize is given to only one toy a year and, as far as I know, no other puzzle has ever received it.

In 1981, the New York Museum of Modern Art included the cube in its collection of design and architecture.

The cube has found its way to millions of people, clubs have been established and competitions have been organized in dozens of countries all over the world. Hundreds of articles and at least 60 books have been published about it, showing that the craze for the cube is conquering the whole world. An international cubing championship has been organized.

I need hardly say that I could never have imagined that the cube would enjoy such enormous popularity and international acclaim and even now I can hardly believe it.

2

THE ART OF CUBING

Tamás Varga

2.1 Introduction

The main objective of cubing is to be able to bring back the cube's original one-side/one-colour pattern from any other pattern. (We shall call this original pattern 'start'.) Alternatively one can begin at start and produce other given patterns. The two tasks can also be combined. You can start with a given pattern, and aim to reach another given pattern.

There are many other tasks which can be set, some of which can be used as preliminary steps towards unscrambling the cube. You will find a list of them at the end of this section, on p. 21.

The greater part of the present chapter deals with the problem of unscrambling the cube—our main objective! We shall describe a few simple techniques which can be fitted together into easily learnt formulae for solving the cube.

In addition, these techniques will help you understand how the cube works and help you solve various other problems.

Undoubtedly the greatest source of satisfaction will come from being able to work out for yourself how to solve the cube. If you are one of those who hate the easy way out then do not read further. But be warned, it is not as easy as it looks!

The rest of you can read on and be guided by the following exercises, which we shall call *organized tours.*

The tours always begin with the cube in its start position. Often two, or sometimes three, moves will be repeated. The everchanging patterns will show you at a glance how your movements affect the cube. Then you can decide if there are other ways in which you can use a technique. These techniques will sometimes have to be combined or even modified if they are to help us unscramble the cube.

The simple repetitive movements require a minimum of mental effort to learn and a minimum of manual dexterity to perform them quickly. The illustrations will help you visualize both the processes and their results. Try to turn each formula into a slogan, so that you do not have to rely on your visual memory or on the motor memory which controls your finger movements.

So how do we get back to our starting position at the end of a tour?

This is no problem; you will automatically return to it because the same moves are repeated again and again. But a false move will spoil everything. The colours will immediately get out of sequence, and you have not yet mastered the technique of solving the cube. So what should you do?

Regard it as an excellent opportunity to do some experimenting. (The exercises at the end of this section lead into precisely this kind of activity.)

But let us suppose you want to try out some of the organized tours, and there is nobody to restore your cube for you. You can either rearrange the stickers, or—better still—take the cube apart and reassemble it. Start by selecting one of the edge-pieces and rotate one of the faces it lies on through 45°. In this position the edge-piece can easily be removed with a knife or screwdriver. When reassembling the cube, the final step will be just the opposite of the way you began: putting back a edge-piece into a face rotated through 45°.

Incidentally, although corner-pieces and edge-pieces give the impression of being complete cubes themselves, taking the cube apart reveals that none of them are actually cubes. (From here on the words 'corner', 'edge', and 'centre' will be used for short.) Figure 2.1 indicates some other commonly used names. However you look at the cube, you can see at the most the three coloured faces. When you look at the cube from one corner, this we will call the *main corner.*

The 12 edges can be moved freely and so can the eight corners, but not the centres, they remain fixed to each other. If, for instance, the blue centre is opposite the green centre on a cube, then it remains so the whole time, no matter how the faces of the cube are rotated. As a preliminary to working with the cube it is worth noting which colour is on the top and which colour is on the front.

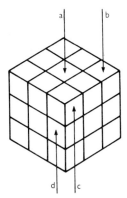

Figure 2.1 The three different kinds of small cube with one, two or three coloured faces
    (a) centre-piece, face centre-piece
    (b) corner-piece, corner-cube, corner
    (c) main corner
    (d) edge-piece, edge cube, edge

The illustrations in this chapter are based on an unscrambled cube with blue on top, red in front and yellow at right, where the directions are arranged as shown in Figure 2.11. The faces opposite the above are green, orange, and white respectively. The colours appearing on opposite faces at start will be called 'opposite colours'.

A corner or an edge may or may not be positioned correctly and even if it is in its correct position, it may or may not be properly oriented. There is only one way for an edge to be incorrectly oriented when it is correctly positioned (Figure 2.2b). There are

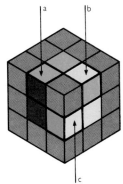

Figure 2.2 (a) Edge correctly placed and oriented
    (b) Edge correctly placed but wrongly oriented
    (c) Alien edge

two ways for a corner to be incorrectly oriented—when it is correctly positioned either it is twisted clockwise (Figure 2.3b) or anticlockwise. In order to decide which of these is the case you simply have to look at the centres of the three faces surrounding the corner.

Practice will enable you to extract considerably more information from what you see with or without moving the cube. An edge or a corner is all right if it is both positioned and oriented as in Figure 2.2a and Figure 2.3a.

 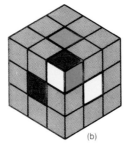 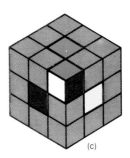

(a) (b) (c)

Figure 2.3 (a) Corner correctly placed and oriented
(b) Corner correctly placed, but twisted clockwise
(c) Alien corner

One more preliminary remark: whatever the merits of the strategy developed in this chapter, it is certainly not meant for championship contenders. More efficient, but also more sophisticated and exacting strategies can be learned from, or based on, the information in the next chapter.

Exercises

1. By turning the faces of the cube make each of the six colours appear on one face, then on two faces . . . then on as many faces as you can.

2. Reduce the number of colours on a face to two; then to one.

3. Make the number of colours on each face less than six; less than five; less than four; less than three. The last step is not easy if you start with a scrambled cube. If you scramble the colours by half turns (i.e. 180° rotations) from start you will find that you cannot get three colours on any face.

4. Add up the number of colours on each face. Try to make the

answer as large as you can. (If you can make it 36 it means you have solved the most difficult part of exercise 1.)

5. Try to decrease the number of colours on each face. Reducing the total of these numbers to 6 means you have solved the cube.

6. Make one of the colours, such as the red, appear on each of the faces.

7. Make two of the colours, such as red and blue, appear on each of the faces. Do the same with three colours, then four, then five. Making all six colours appear on each face means that you have again solved the most difficult part of exercise 1.

8. Move the four corner-pieces which have blue facelets on to one face, around the blue centre. Then none of the corners on the opposite face will bear a blue facelet; each will have a green facelet. (A convention: pieces with a blue, green, etc. facelet will be called blue, green, etc. pieces.) Note that a piece (corner or edge) can be blue and red at the same time, but it cannot be blue and green at the same time if the cube has its blue centre opposite its green centre.

9. Move the blue corners *and* the blue edges so that they are around the blue centre.

10. As above, but try to get as many of the blue facelets as you can onto the plane of the blue centre. If you succeed in bringing all of them into the same plane, then then difficult part of exercise 2 will have been solved: one face, one colour.

11. A simplified version of the above. Bring the four edges with blue facelets around the blue centre, forming a blue plus (Figure 2.4).

12. A more difficult version. Try to get each of the four blue edges correctly positioned and oriented (Figure 2.5).

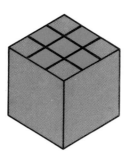

Figure 2.4   Blue +

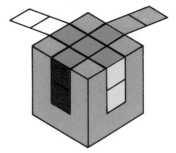

Figure 2.5   Blue   edges   correctly positioned and oriented

13. Bring the four corners with blue facelets around the blue centre, forming a blue × (Figure 2.6).

Figure 2.6 Blue ×

14. Make each of these four corners correctly positioned and oriented. The letter 'V' will appear in one colour on each of the four faces adjacent to the blue one (Figure 2.7).

Figure 2.7 Blue corners correctly positioned and oriented

15. Make each of the four blue corners and four blue edges correct (Figure 2.8).

Figure 2.8 The blue layer is correct

Having solved exercise 15 amounts to solving one layer. Most of the published and practised strategies for solving the whole cube start with one layer. Having achieved this means much more than reducing one face to a single colour as in the difficult parts of exercise 2 and 10. Solving one face is an interesting exercise in itself, but it is only a beginning if your goal is to have each of the faces of one colour.

2.2   A 'mini-tour': half turns of two faces repeated

Start with the cube at start. Begin turning two adjacent faces alternately, for example, the front and the right faces, (see Figure 2.11) but *always with half turns* (i.e. 180° rotations). The monochrome faces will change to show two colours, which are opposite colours. The double arrows in the upper part of Figure 2.9 indicate that the faces containing the arrowheads should be rotated 180°. Half turns can be performed either forwards or backwards. The symmetry of the double arrows is a way of expressing this idea.

Repeat the half turns. You will see the same colours again, but the patterns will be different.

After doing the same thing a third time, four of the faces will again become a single colour. Two colours will appear on the top and bottom faces — two facelets will be different from the rest. The lower right part of Figure 2.9 shows the result.

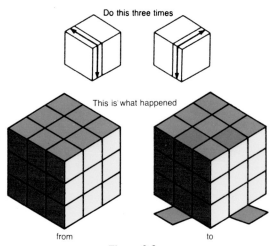

Figure 2.9

One's first impression is that two pairs of facelets were exchanged, two blue ones for two green ones, and that nothing else happened. Actually two red ones and two yellow ones were also exchanged. In fact the two pairs of edges bearing these four facelets changed places. Figure 2.10 shows this. Two of the four double arrows would suffice to define the movements. If the new position of one facelet of an edge is known — or, for that matter, of a corner — then the movement of the other(s) can be deduced.

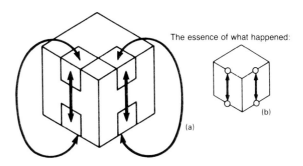

The essence of what happened:

(a)

(b)

Figure 2.10 (a) Two pairs of edges are swapped. (b) Another way of showing this.

Four of the edges moved; eight of them remained in place as did the eight corners. (We mean remained in place in the sense that only the end result of a sequence of moves matters. The moves in between are immaterial.)

Swapping two pairs of edges as in Figure 2.10 is one of the basic skills of a cubist. These moves should be recalled by one's tactile memory and performed automatically, leaving the visual memory free to concentrate on what is expected to happen. But even a skilled cubist's memory can be supported by visually recalling the sequence of moves (as in Figure 2.10) before performing a less frequently used, less automatic process. Illustrations also enable us to share with others any new things we learn.

Illustrations help the imagination as well, but *literal notation* also has its advantages. It can be used to record the moves quickly and concisely. David Singmaster's notation is the most widely known in the English-speaking world (Figure 2.11).

U=up	D=down
F=front	B=back
L=left	R=right

Figure 2.11 T = Top   D = Down
              F = Front   B = Black
              L = Left   R = Right

In this book we keep five of these six letters, changing U to T (for top) to make it a consonant like the other five; the reason for this will become clear in a moment.

As well as denoting the six faces (or outer layers, each consisting of nine components) of the cube, these letters also indicate clockwise quarter turns of these faces as shown in Figure 2.12. Clockwise on a face is always determined as one looks directly at the face. A

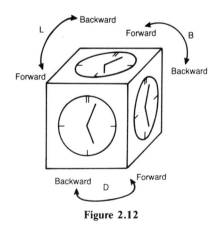

Figure 2.12

half turn being equal to two quarter turns, the process which we have just carried out can be recorded as:

$$FFRRFFRRFFRR$$

Writing the letters close to each other from left to right indicates the sequence of the moves denoted by the letters. In algebra, writing two letters close to each other indicates multiplication. Combining turns in a sequence may also be considered to be multiplication in a very general sense, and so we shall use the power notation for products of equal factors. Thus the above sequence becomes:

$$F^2R^2F^2R^2F^2R^2 \text{ or } (F^2R^2)^3$$

This is usually pronounced *F squared R squared F squared R squared F squared R squared* or *the cube of F squared R squared*.

The following is a shorter way of reading this:

firi firi firi or *three firis*.

Here *i* means twice; two quarter turns make a half turn; we use *i* since it is the first vowel in tw*i*ce.

Two anticlockwise quarter turns produce the same effect as two clockwise ones. One anticlockwise *quarter* turn, however, is different from a clockwise quarter turn. In the usual mathematical notation, the moves opposite to T,D,F,B,L,R, are $T^{-1}, D^{-1}, F^{-1}, B^{-1}, L^{-1}, R^{-1}$, respectively (read *T to the power of minus one, D to the power of minus one ...* or *T inverse, ...*). David Singmaster adopted the shorter 'prime notation' instead. This turns the above to T', D', F', B', L', R', (*T prime, D prime ...*). We suggest you read them:

ta, da, fa, ba, la, ra.

The vowel *a* is the first vowel of *a*nticlockwise or *ba*ckward.

In a similar way, the clockwise quarter turns of faces denoted as T,D,F,B,L,R, may be read as

to, do, fo, bo, lo, ro

respectively, *o* being the first vowel of f*o*rward, or of cl*o*ckwise.

Now let us look at the F^2R^2 (*firi*) moves when you carry on after the third. What will happen?

It is easy to guess: three F^2R^2s (*firis*) having swapped two pairs of edges, the next three will swap them again, putting them back where

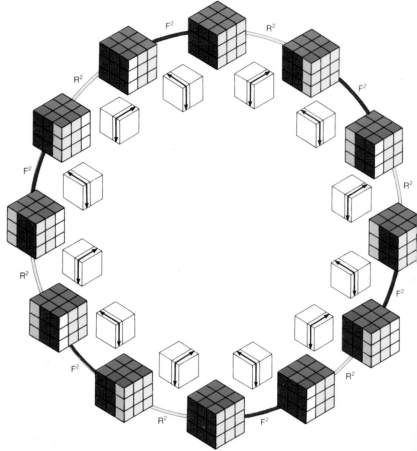

Figure 2.13   Half turns of two adjacent faces generate 12 different states of the cube

they were in the first place. Figure 2.13 shows every stage of the round trip—$(F^2R^2)^6$ or (six *firis*).

Just as repeating F^2R^2 brings us back to our starting point after a while, repeating *any sequence of moves* does the same. This may happen after just a few repetitions of the move sequence, but you may have to do the same thing as many as 1260 times. This is the maximum number of moves for returning to your starting place. Sequences far shorter than this can confuse you and lead to mistakes. So repeating the same moves until you get back to the starting position is not always the best policy. Often it is worth returning by a short cut before you make a false move. Recording the moves — using either the formulae or the syllable notation — will be useful for getting you back. In order to undo F^2RF^2R (two *firos*), you start with the opposite of your last move. The opposite of R is R′. Then you do the opposite of F^2, which is F^2 since half turns arrive at the same final position whether they are made in a clockwise or an anticlockwise direction. And so on. The opposite to F^2RF^2R or *firo firo* will be $R′F^2R′F^2$ or *rafi rafi*; more concisely, the opposite to $(F^2R)^2$ or two *firos* is $(R′F^2)^2$ or two *rafis*.

Another example, $RF′T \cdot L^2B′$ (*rofato liba*) can be undone by $BL^2 \cdot T′FR′$ (*boli tafora*). Longer words may conveniently be broken down into shorter ones, the parts being separated by a dot, $\cdot$, in the formula or by a space in the syllabic notation. If you have any doubt as to how to invert a word just think of the sequence of moves to be undone. You will find it quite natural that both the order of the moves will be reversed and every move is itself reversed. This is also the case in everyday actions. For instance, pulling on your socks and putting on your shoes has as its inverse: first taking off your shoes, then pulling off your socks.

Is the inverse order really necessary in every case? Yes it is, unless the order of the operations makes no difference. Putting on your shoes and then pulling on your socks has quite a different effect from doing the same thing in the inverse order; but whether you put your right sock or your left one on first, the result is the same, and when removing them at night you do not have to do the inverse of what you did in the morning.

Now the question is whether there is anything analogous to this in turning the faces of your cube. Have you a similar freedom when you make the inverse of a move sequence? (This is problem 0.)

The rest of the problems, at the end of this section, will be about turning two adjacent faces: the front and the right face. You can choose any two adjacent faces and then hold the cube in such a way that the first of these faces is in front and the second to your right; therefore choosing F and R does not imply any restriction.

Do not forget to count the number of times you perform the two moves, saying them to yourself as you do them: *first time, second time. . . .* Let us start an organized tour. Do the sequence F^2R (*firo*) a few times.

After you have done it *six times* you will find the corners correct. Not just four of them as shown in Figure 2.7, but all eight as indicated in Figure 2.14. Three of the four edge facelets on each

Figure 2.14 Corners all correct: each face shows an × of one colour (each of the remaining four squares may or may not be of the same colour)

face also bear the colour of the corner facelets. Let us continue with the F²Rs (*firo*s). After having done ten of them, you will find the edges correct, just like the blue ones in Figure 2.5. On each face you will see + signs as shown in Figure 2.15. (Again, some of the corner facelets are also of the same colour.) After $(F^2R)^{12}$ or 12 *firo*s, the corners are correct again. The same happens after 18 *firo*s. After $(F^2R)^{20}$ or 20 *firo*s the edges are all correct again.

Figure 2.15 Edges all correct: each face contains a + (each of the remaining four squares may or may not be of the same colour)

The pattern which emerges is this: after every six *firo*s all the corners are back in place; after every ten *firo*s so are the edges. The point at which the edges are both back in place thus bringing the cube back to its start, will be where these two sequences meet:

$$6, 12, 18, 24, \ldots$$
$$10, 20, 30, 40, \ldots$$

This happens first at 30, then at 60, etc. The first sequence consists of multiples of six, the second of multiples of ten; 30 is their least common multiple. After $(F^2R)^{30}$ or 30 *firo*s the tour returns to start. If you go on with the *firo*s, you are repeating the same tour.

Exercises

(The solutions are at the end of this chapter.)

1. After how many FR or *foro* turns will the edges all be correct? The corners? The cube as a whole? (The answer to the last question can be deduced from the two previous answers.)

2. The same as exercise 1, but with F'R' or *fara* turns ... ? (The same questions.)

3. The same for FR' or *fora*.

4. The same for F'R or *faro*.

5. Now try FR^2 or *fori*.

6. Try these: F^{2}R' or *fira*, F'R^2 or *fari*.

7. If you are ready to work really hard, then do F^{2}R or *firo* just once, and try to work out from what you see why the numbers 6, 10, and 30 occurred when experimenting with the *firos* above. (Observe what happened to any one of the cubes which moved. Where did it go to? Where did the one it replaced go? And so on. Think of what will happen if you do the same thing again. The insight you will gain will not be restricted just to repeated *firos*!)

8. So far our only concern has been to see how many *foros* etc. were needed to get the corners or the edges or both back into correct position and orientation. Let us go back to the first six exercises, asking ourselves this time a different question: how many *foros* etc. will correctly position all the corners, *even if twisted,* i.e. incorrectly oriented. (Look at the cube in the middle of Figure 2.3 to see how you can decide if a corner is positioned but twisted.) Which corners will be twisted and in which way in each case? (Ignore the edges.)

9. Do the same kind of experiment with these sequences:

(a) FTR (*fotoro*) and (b) R'T'F' (*ratafa*),

(c) FT^{2}R (*fotiro*) and (d) R'T^{2}F' (*ratifa*).

2.3 Twists

Those who solved problem 8 of the previous section can twist six corners forwards or six corners backwards. They can also twist three corners forwards and another three backwards, and in two different ways. The six corners twisted are always the same: those which lie on the front and right faces. Clearly, if only these faces are turned, then the two corners not on these faces remain unmoved.

Those who have solved problem 9 can twist three corners forwards or three corners backwards, keeping five corners unchanged.

Twisting corners is a useful skill, but how can it be fitted into a strategy which will solve the cube?

One strategy is this: first position the corners, then we twist corners using this skill and then we position and flip the edges, keeping the corners unchanged. The rationale for this is that twisting the corners as we have just learnt scrambles the edges so that the edge restoration had to be postponed. On the other hand we should not begin twisting the corners until they are positioned when it becomes clear what twists are actually needed. (Although we have not yet seen how to carry out all these stages, the study of F^2R^2 gives us hope that we will be able to do them.)

There are many other ways of solving the cube. For instance, solving it layer by layer, in which case attention has to be paid to keeping the solved layers unchanged. But our system will be different. First we position the eight corners, orient them, then we position the twelve edges, and finally orient them.

Our approach has led us to the second stage first. Let us keep on with it for a little while longer.

Let us assume that all corners are in the right place, but some are twisted forwards, some backwards and some are correctly orientated. We wish to orient all of them correctly. Would our present knowledge of move sequences be sufficient for the purpose? We do not often need to twist six corners simultaneously in the same direction, (problem 1, section 2.2), or turn three one way and three the other way, as is done by $(FR')^3$ (three *foras*) or by $(F'R)^3$, (three *faros*, as shown in Figures 2.16 and 2.17.

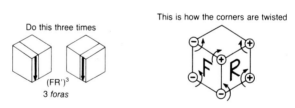

Figure 2.16 Twisting three corners clockwise, three anticlockwise

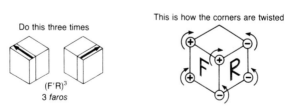

Figure 2.17 Twisting three corners clockwise, three anticlockwise, in a different way from Figure 2.16

Techniques which twist only three corners are more promising, such as $(FT^2R)^3$ (three *fotiros*) (anticlockwise) and $(R'T^2F')^3$ three *ratifa*s (clockwise); (see Figures 2.18 and 2.19, respectively).

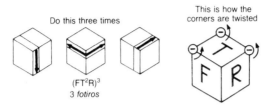

Figure 2.18   Three corners twisted anticlockwise

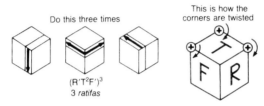

Figure 2.19   Three corners twisted clockwise

If you bear in mind the multitude of ways in which the corners can be twisted, then you will find your present state of knowledge somewhat scanty. You may want to twist three corners independently: not all the same way, but two of them one way and one the other; or you may want to twist two corners, not three — both clockwise, both anticlockwise or in opposite directions. And what about twisting just one corner, either clockwise or anticlockwise? Also, are we limited to twisting three corners the same way only if they are all in the same layer?

In fact this is wishful thinking. Quite a few of these cases are not possible. If the corners are to keep their positions, one of them on its own cannot be twisted either clockwise or anticlockwise; two cannot be twisted both clockwise or both anticlockwise; three can be twisted *only* if all of them are twisted in the same sense. (For the moment, we only consider corners which are properly placed; this restriction will be removed when we get to Section 4.) Fortunately, the limitations do not make the solution of the cube more difficult. It is not that we *need* certain kinds of twists and *cannot* perform them — we *do not need* them, because the situations where we would need them simply cannot occur. For example, it is impossible to have just one corner twisted while everything else (or, for that matter, the other corners) is correct. Nor is it possible to have just two corners twisted the same way. If you see two corners twisted the same way, and all the corners are in their correct positions, then you can be sure there will be at least one more corner twisted — unless you have been tricked by somebody who has reassembled the cube improperly or changed its stickers!

Bearing this in mind you may find it less of a suprise that your present knowledge of twisting corners allows you to orient them in any situation. Remember that two opposite twists of a corner cancel each other out while two twists in the same direction are the same as one in the other direction. We can represent this symbolically by

$$+ - = - + = 0$$
$$+ + = -$$
$$- - = +$$

In fact, just *one* of the two triple twists is sufficient to generate every possible set of twists. First, twisting three corners forwards *twice* is equivalent to twisting them backwards once. Second, two corners can therefore be twisted in opposite directions (see Figures 2.20 and 2.21). Third, combining such twists of two corners produces every set of twists which is permitted by our previous discussion.

If we study and practise more than just one process of twisting corners, it will enable us to restore the cube more efficiently. The process in Figure 2.21 with its $6 + 6$ moves is more efficient and, hence, preferable to the one in Figure 2.20 which takes $9 + 9$ moves. (Moving the cube itself is not counted.)

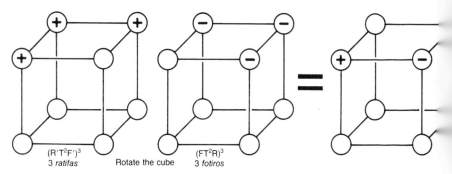

$(R'T^2F')^3$ $(FT^2R)^3$
3 *ratifas*   Rotate the cube   3 *fotiros*

Figure 2.20   Twisting two corners in 18 turns

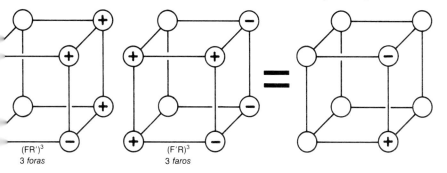

Figure 2.21 Twisting two corners in 12 turns

The shorter of the two processes can be transformed so as to twist the two front top corners. You simply turn the front and the top layer in the same way you turned the front and the right layer previously (Figure 2.22).

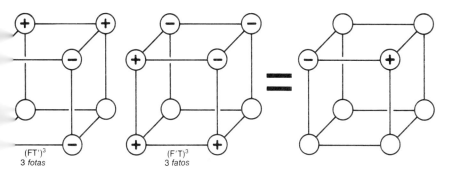

Figure 2.22 Twisting two top-layer corners in 12 turns

It is now time to give a comprehensive answer to the question: what are the possible patterns of corner twists which leave the corners in place? By combining twists as in Figures 2.20 and 2.21 you will soon realize that you can get the following.

(a) Twisting one, two, three, or four corners clockwise and the same number of corners anticlockwise:

$$+-, \quad ++--, \quad +++---, \quad ++++----.$$

(b) Twisting three more or three less clockwise than anticlockwise:

$$+++, \quad ++++-, \quad +++++--,$$
$$---, \quad ----+, \quad -----++.$$

(c) Twisting six more or six less clockwise than anticlockwise:

$$++++++, \quad ++++++++-,$$
$$------, \quad --------+,$$

Mathematically speaking the rule is this: by replacing the $+$ symbols by $+1$, the $-$ symbols by -1, the sum of the numbers should be a multiple of 3. (The multiples of 3 are 0, 3, -3, 6, -6, ...) This statement is consistent with the one that twisting three corners in the same sense can generate every set of twists. (Such a triple twist changes the sum by a multiple of three. At start the sum is 0, which is a multiple of three.)

The rule can be worded less mathematically. Think of twisting one corner, without worrying about what happens to the other corners. Twisting it three times in the same sense restores its previous orientation. So does four twists clockwise and one twist anticlockwise, etc. The twists you can do to any number of corners are precisely those which, if applied to one corner alone, would eventually *do nothing* to it. Before the exercises, here is a practical point related to twisting three corners in a layer, for example, take the top layer. Suppose the corners here are in place, but that three of them need twisting to make them correct. Figure 2.23 illustrates this: one of the corners is correct, and the blue facelets of the other three, which should be on the top, form a windmill-like configuration. You cannot actually see this (except perhaps by means of a curved

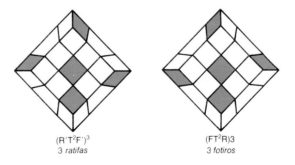

$(R'T^2F')^3$
3 *ratifas*

$(FT^2R)3$
3 *fotiros*

mirror arrangement). Yet if you look at the cube from the top and rotate it slowly in order to see the layer around the top face, then you will get an image which is the same as that shown in the picture.

Exercises

1. The corners are all in place, but two facelets on the front face, as in Figure 2.24, need to be moved to the top. What should you do?

Figure 2.24

2. The inverse of the above: two facelets have to be moved from the top to the front (Figure 2.25).

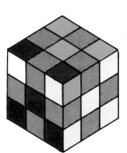

Figure 2.25

3. Every corner of a cube is in place, but those in the top layer are all twisted. (Those in the down layer are correctly oriented.) It follows that two are twisted clockwise and two anticlockwise. Those twisted the same way can be either (a) opposite, or (b) adjacent (Figure 2.26). Try to find ways of orienting the corners in both cases.

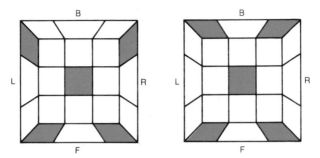

Figure 2.26   The four upper corners twisted in two different ways

4. The following techniques are not exactly repetitive, but are nearly so. What is their effect on the cube? What have they to do with the previous problem?

(a) TF^2 · T^{2}F · T^{2}F^2 (*tofi tifo tifi*),
(b) R′(T′F′ · TF)^{2}R (*ra tafa tofo tafa tofo ro*).

(The dots are used to break up the notation.)

5. These, too, are almost repetitions. Try them! In what way is the second related to the first?

(a) FT · F″T · FT^2 · F′T^2 (*foto fato foti fati*),
(b) T^{2}F · T^{2}F′ · T′F · T′F′ (*tifo tifa tafo tafa*).

6. Now this one: F″T′ · FT′ · F′T^2 · FT^2 (*fata fota fati foti*). In what way is this one related to 5(a)? (Now compare both sequences and their effects.)

7. Similarly, compare the previously examined sequences (FT^{2}R)^3, three *fotiro*s, and (R′T^{2}F′)^3, three *ratifa*s. More about such transformations can be found in section 2.13.)

2.4   Conjugation: a useful ploy

By learning new sequences, many superfluous moves can be eliminated. By using conjugation, the learning of many processes can be made more economical. The idea is that the scope of a known process can be extended by conjugation to many cases not previously covered. For example, we have learned how to twist two adjacent corners. How can one twist two corners, if they are opposite each other on a face? Specifically, how can we move the blue facelets on the left and right of Figure 2.27 (on the front and right faces in our standard notation).

The trick is this: make the two corners adjacent by turning a layer, for example, by R′ (*ra*). Now they can be twisted by (FT′)³ (F′T)³ or three *fota*s, three *fato*s (Figures 2.28 and 2.29). Only one of the blue facelets is on the top and some corners are displaced. But if you undo the introductory R′ (*ra*) move by R (*ro*), then the corners will be in place again, with two of them twisted as desired.

How can we do this . . . if we know how to do this?

Figure 2.28   The unknown situation can be changed to a known one

Figure 2.27

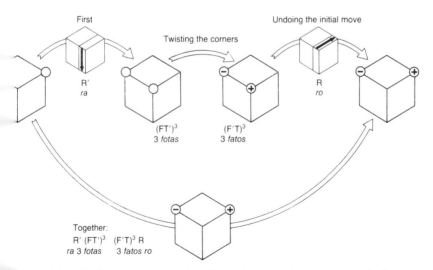

First

Twisting the corners

Undoing the initial move

R′
ra

(FT′)³
3 *fotas*

(F′T)³
3 *fatos*

R
ro

Together:
R′ (FT′)³   (F′T)³ R
ra 3 *fotas*   3 *fatos ro*

Figure 2.29   If adjacent corners can be twisted, then opposite corners can also be twisted

It can also happen that the two corners are not in one layer as shown in Figure 2.30. There are just these three arrangements of two corners.

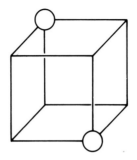

Figure 2.30   Two corners which are farthest apart

There are also just three relative positions of three corners (see Figure 2.31):

(a) all in one face;
(b) two adjacent in one face;
(c) non adjacent.

In case (a) we know how to twist them, if they are placed as in Figure 2.31, either all clockwise — by $(R'T^2F')^3$, three *ratifa*s — or all anticlockwise — by $(FT^2R)^3$, three *fotiro*s.

But what do we do in cases (b) and (c) — problems 2 and 3 below, respectively?

Problem 4 presents a chance to apply conjugation to swapping edges. The main idea of conjugation is that one or more introductory moves are used to create a situation which we already know

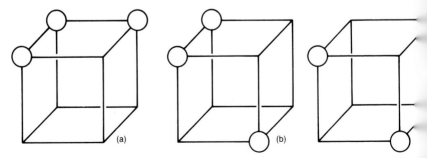

Figure 2.31   The three possible arrangements of three corners: (a) all in one face; (b) two adjacent; (c) non adjacent

how to resolve. These introductory moves have to be undone at the end, according to the 'socks-and-shoes' rule, that is, doing the inverse of each move in the inverse order.

Problems

1. Twist two corners which lie on a diagonal of the cube. Record the moves for two corners located as in Figure 2.30.
2. Twist three corners of a cube and record the moves for corners placed as in Figure 2.31b.
3. Do the same for Figure 2.31c.
4. Swap two pairs of edges in the positions shown in Figure 2.32. (Each can be reduced to the process $(F^2R^2)^3$, three *firis*, of section 2.2.)

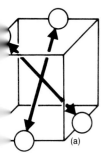 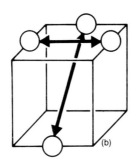 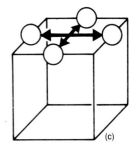

Figure 2.32

2.5 How to position the corners

Look at Figure 2.20 in section 2.3. In order to twist *two* corners, we combined the twisting of *three* corners in two different ways. Between these twists, the cube itself had to be turned. This move has not been recorded so far, either graphically or symbolically, as have other moves. This gap will now be filled. Consider Figure 2.33, where the right diagram shows a movement of the whole cube.

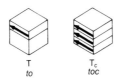

Figure 2.33 T means turning the top face clockwise; T_c means turning the whole cube in the same direction

The triple arrows to the right indicate that all three layers, that is, the cube as a whole, is turned with the top layer, to the left. This is written T_c and subscript c stands for *cube* (Figure 2.33).

What was hinted at in Figure 2.20 was not this move, but the opposite quarter turn around the vertical axis, namely T'_c, or *tac*. A final T_c or *toc*, missing in Figure 2.20, completes this process; this provides a way of getting the twisted corners back to where they are in the right-hand cube of Figure 2.20. (So this is also a conjugation. The introductory move has to be undone.)

T^2_c, or *tic*, means a half turn around the vertical axis. The following can also be interpreted analogously: F_c, F'_c, F^2_c (*foc, fac, fic*); R_c, R'_c, R^2_c (*roc, rac, ric*). No others are necessary since L_c is the same as R'_c, etc.

Having introduced these moves, the scope for our exploratory tours becomes much wider. Moving only faces led us to repeat triples of moves in order to get interesting results without too many repetitions. Now we can return to repeating pairs of moves such as these (Figure 2.34).

$$FT_c, \quad FT'_c, \quad FT^2_c, \quad F^2T_c, \quad F^2T'_c, \quad F^2T^2_c$$
$$\textit{fotoc} \quad \textit{fotac} \quad \textit{fotic} \quad \textit{fitoc} \quad \textit{fitac} \quad \textit{fitic}$$

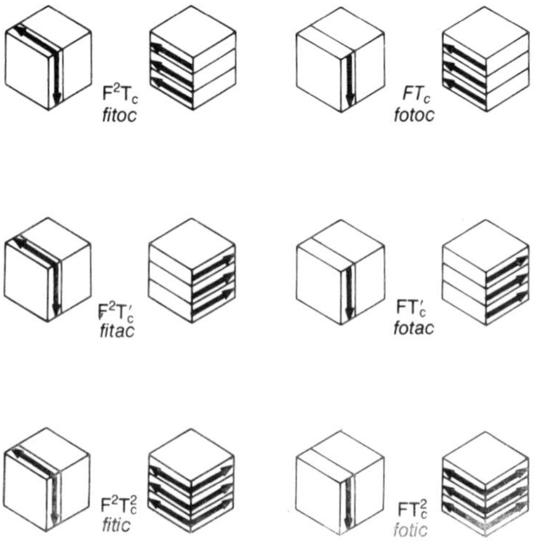

F^2T_c fitoc

FT_c fotoc

$F^2T'_c$ fitac

FT'_c fotac

$F^2T^2_c$ fitic

FT^2_c fotic

Figure 2.34

Ignore the edges for now. Do not expect the corners to keep their places as before. The aim now is to move some of them but only a few as we want to see what is happening. (These are problems for you to do, but the solutions are not given. We recommend you work on them before proceeding further with the book.)

The last four of these six moves do not produce interesting results. However, the first two do and after only four repetitions: both $(FT_c)^4$, four *fotocs* and $(FT'_c)^4$, four *fotacs*, cycle three corners while keeping five in place. (We shall call this a three-cycle or a *tricycle*.) $(FT'_c)^4$ has the advantage of moving three top layer corners (keeping only the main corner.) This is shown more precisely in Figure 2.35, which shows the moves, the effect on the entire cube (shown from front and back) and, schematically, the effect on just the corners.

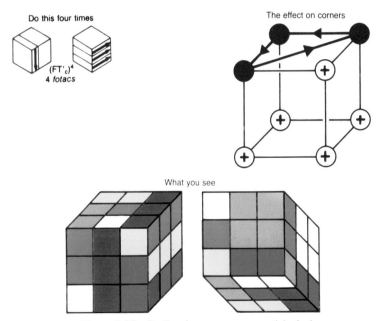

Figure 2.35 Cycling three top corners anticlockwise

The three corners are cycled anticlockwise. (The remaining five are twisted, but we are not concerned with that. In the strategy we are developing to solve the *scrambled* cube, orienting the corners comes after positioning them. Before that stage, twisting them any way can be either good or bad; on average it is neither good nor bad.)

Now try to find a process which cycles the same three corners *clockwise*! Looking at Figure 2.20, we see two different relationships between $(FT^2R)^3$, three *fotiro*s, and $(R'T^2F')^3$, three *ratifa*s, namely that they are both the inverse and the mirror image of the other. We can supply these relationships to $(FT_c)^4$, four *fotoc*s, and $(FT'_c)^4$, four *fotac*s, to find two solutions.

By inversion	By reflection
$(T_cF')^4$	$(R'T_c)^4$
four *tocfa*s	four *ratoc*s

If you do both, you will find that they are basically the same. Though written differently, inversion and reflection produce the same effect. Now let us look at the second variant more closely (Figure 2.36).

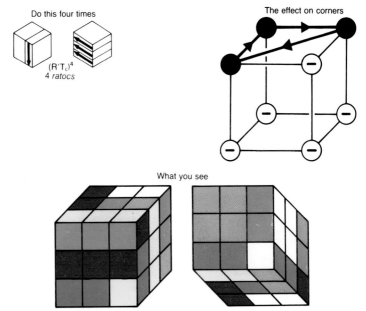

Figure 2.36　Cycling three top corners clockwise

Is there any advantage in moving the cube, apart from turning its faces, during a process? Could we not perform the same process without moving the cube? We could — the former one by F L B R, its mirror image by R′B′L′F′. Both become shorter: four moves each time, instead of four times two, or eight. (By the way, inversion and reflection coincide exactly this time.)

However, this advantage is only in the formulae. Use a stopwatch to time 10 or 20 FLBRs or *foloboros* and the same number of $(FT'_c)^4$ or four *fotac* sequences (or slightly different ones which you find more convenient, cycling other corners), and see the difference. A contest between two people each using both methods is still more convincing. In this case the net result is that eight moves can be done in less time than four. The repetitions actually make the eight moves easier for the fingers as well as the mind. Changing one's physical grasp on the cube so as to avoid mistakes does take more time and mental energy than doing the same four turns and moving the cube in between. [With practice, one carries out both sequences in the same way and the question is whether one formula is more convenient to write or read or learn or understand than the other; this varies with the problem and the context — DS.]

Now take a cube in its unscrambled state and cycle three corners, for example, by $(FT'_c)^4$, four *fotacs*. Now find a single turn which will get six rather than five corners in place. In other words, the goal is to get two of the six corners swapped.

Figure 2.37 illustrates the solution and helps you see how it works. $(FT'_c)^4$, four *fotacs*, moves three corners, *a, d, c*. T, *to*, moves

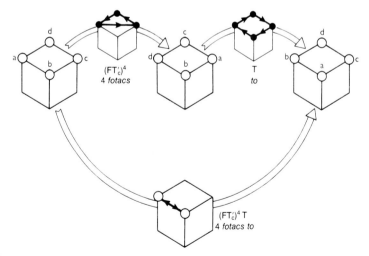

Figure 2.37 Cycling three corners plus turning a face comes to the same as swapping two corners

a fourth, *b*, but returns two, *d* and *c*. The net result is swapping *a* and *b* by $(FT'_c)^4T$, four *fotacs* and *to*. The whole story is shown in Figure 2.38.

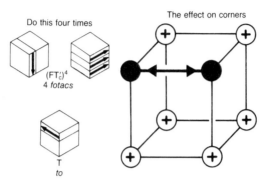

Figure 2.38   Swapping the top front corners

Two adjacent corners can always be swapped this way, holding the cube so as to have them in the front top positions. If the two corners are not adjacent, then one of them can be moved to make them so (exercise 1).

Now that you can swap any two corners, there is an easy way of positioning all the corners.

Look at a corner not yet in place. Find the one in the first one's place (in its *cubicle*). Swapping them does not spoil anything. (The second corner is in the cubicle of the first, and therefore it is not yet in place.) The number of corners in place will be increased by one, or possibly by two, if they were in each other's cubicles. Go on in this way. In the worst cases you will need seven swaps. The last two corners must go home simultaneously.

In the next section a quicker way of positioning the corners will be developed. Those of you who are not interested in such short cuts may skip it. But do not skip the exercises at the end of section 2.6; solving them will make you ready for the next two stages which solve the edges.

Exercises

1. Find processes for swapping two corners:

(a) both on the same face, but not adjacent;
(b) not on the same face.

2. Find a tri-cycle of corners not all on the same face

(a) with two adjacent;
(b) with no two adjacent.

3. On the left side of Figure 2.39 a cube is shown with its corners *a, b, c, d, e, f, g,* and *h* scrambled. The cube furthest right in the second row (cube 7) indicates the correct order. Find processes to position the corners (a tri-cycle is shown to start with).

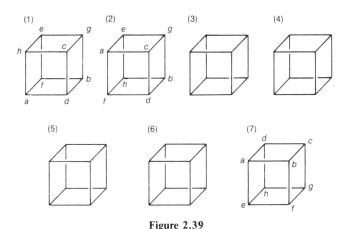

Figure 2.39

2.6 How to position corners quickly

It was easy to *describe* a general method of positioning the corners using only swaps by $(FT'_c)^4 T$, four *fotacs* and a *to*. However, it is easier to *perform* the positioning using tri-cycles of corners: clockwise $(R'T_c)^4$, four *ratocs*; anticlockwise $(FT'_c)^4$, four *fotacs*.

Let us outline how this can be done. Choose a colour, say blue. Get the blue corners into one layer. (Remember that a blue corner means a corner with one blue facelet.) But how? We will come back to this point later. If the centre of the face is not blue, then the blue centre can be moved there and we will also describe this later. So we now assume the blue corners are in the blue face. Hold the cube with the blue centre facing upwards. The blue corners are in the top layer, the non-blue (in our case, green) corners are down. If, by chance, every blue corner is in its cubicle or can be moved there by turning the top layer, then we can turn the cube over and start positioning the green corners. If this is not the case, one or two corners can be positioned by turning the top layer leaving three or two corners in position. Three corners can be positioned by $(FT'_c)^4$, four *fotac*s, or $(R'T_c)^4$, four *ratoc*s. Two adjacent corners can be positioned by $(FT'_c)^4T$, four *fotac*s *to*. If we have two opposite corners, then the latter process will work if it is preceded and followed by a move; in other words, if we apply conjugation. Swapping (which is one move longer than a tri-cycle) is actually needed only for two corners. With practice, you can combine these tri-cycles and swaps in many ways. However, deciding among several methods may slow you down. Now you do the same thing on the opposite layer, and you have all the corners positioned.

But how do you get the blue corners into one layer? We shall start from a situation where they *are* in one layer—for example, on the top—and depart from this situation step by step until every possible situation is covered. The steps are reversible: the initial situation was upset by a particular sequence so it can be restored by reversing the sequence.

Imagine the blue corners on the top forming a closed chain. This is represented by the heavy dots and heavy lines in Figure 2.40. The position of the blue facelets on these corners is, for the time being, immaterial.

A quarter turn of any lateral face will open the closed chain. A half turn of any lateral face will break the chain up into two small chains with two corners in each.

Look at the open chain. A clockwise turn of the top face transforms it into a branching chain as shown in Figure 2.40. An anti-clockwise turn breaks off one corner.

Now look at the two small chains. A quarter turn of the front face in any sense breaks it into four separate corners.

It is easy to see that these seven situations cover all possible cases. What you should do with a scrambled cube is hold it so as to match one of them. Then two moves, at most, will create the closed chain situation: the blue corners are in one layer. And hence so are the opposite ones (green in our case). You can, of course, collect the white corners or whichever you find most convenient.

Having the four blue corners on the top, the blue centre may still be elsewhere, on a lateral face, or down. What kind of move will bring it to the top? Try moving a middle layer between two faces.

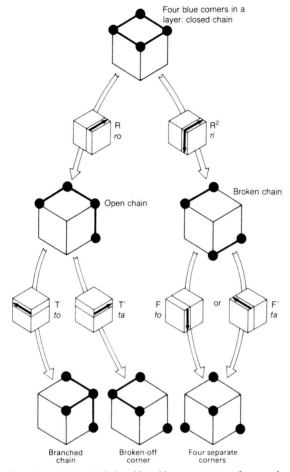

Figure 2.40   How the closed chain of four blue corners on a face can be broken

(This will, at the same time, carry the green centre down.) Figure 2.41 shows three instances of how it can be done, indicated graphically and symbolically. (The index m is for middle layer.) Whether or not you perform this move with a single hand movement or a to and fro movement of faces does not matter: what matters is that our field of exploration has widened again. The problems below will show how.

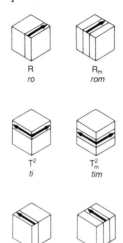

Figure 2.41   Three instances of moving middle layers

Problems

1. Move the middle layers of the cube (this can happen in nine different ways: F_m, T_m, R_m or *fom, tom, rom*; F'_m, T'_m, R'_m or *fam, tam, ram*; F^2_m, T^2_m, R^2_m or *fim, tim, rim*) in any order, as many times as you wish. Look at the patterns which appear on the faces.

   (a) In what way are the colours of the corner facelets restricted?
   (b) How about the colours of the edge facelets?
   (c) How many colours can appear on a face?
   (d) What kind of patterns can appear on a face?

2. Repeat pairs of moves, one of which is a middle layer move. Examine what happens after each repetition. Try these:

   (a) $F^2R^2_m$ (*firim*)
   (b) FR^2_m (*forim*)
   (c) F^2R_m (*firom*)
   (d) FR_m (*forom*)

Which of these could you use to restore the edges, and in what way?

3. Explore what happens if two or three middle layer moves are repeated. Try these:

   (a) $F^2_mR^2_m$ (*fimrim*)

(b) $F_m^2 R_m^2 T_m^2$ (*fimrimtim*)
(c) $F_m^2 T_m^2 R_m^2$ (*fimtimrim*)
(d) $F_m^2 R_m$ (*fimrom*)
(e) $F_m R_m^2$ (*fomrim*)
(f) $F_m R_m$ (*fomrom*)

4. Combine middle layer moves with moving the cube as a whole. Try these:

(a) $F_m^2 T_c^2$ (*fimtic*)
(b) $F_m^2 T_c$ (*fimtoc*)
(c) $F_m T_c$ (*fomtoc*)

2.7 Moving edges while leaving corners fixed

Our explorations of middle-layer moves provide what we need now: processes which affect the edges and leave the corners unchanged.

One way of swapping two pairs of edges is by $(F^2R^2)^3$, three *firis*, the first move we explored in section 2.2. A very similar technique has now been found: $(F^2R_m^2)^2$, two *firims* — similar both in what you do and in what it entails.

Three edges rather than four are moved by $(F^2R_m)^4$, four *firoms*.

How about swapping just two edges? Remember how we found a process to swap two corners. But this cannot be done with the edges fixed! It can be proved (see Chapter 4) that no two cubes, whether corners or edges, can be moved without moving at least one more cube. However, it is possible to swap two corners *and* two edges at the same time while keeping all the other cubes fixed. The minimum number of cubes to leave their places is three; they can be either three corners or three edges. $(F^2R_m)^4$ or four *firoms* illustrates one of these minimal changes very well.

Another kind of minimal change consists of keeping each cube in its place, but at the same time twisting or flipping some of them. We discussed ways of twisting corners in section 2.3 — the minimum number of corners being two. Two is also the minimum number of edges that can be flipped. The general rule for flipping edges is simple: if they all remain in their places, then any even number (2, 4, ... , 12) can be flipped, but an odd number cannot. We have already found a process for flipping four edges, namely $(FR_m)^4$, four *oroms*, but none so far for flipping two. Remember that twisting two corners could be done by combining two processes which twist more than two corners. This implies that there might be a similar procedure for edges.

Our main goal now is to develop a strategy for restoring the edges while leaving corners fixed. This will be based mostly on two processes: cycling three edges and flipping four. This goal is approached through the following exercises.

Problems

1. Stop before completing $(F^2R_m)^4$, four *firoms*, either at $F^2R_mF^2R_m$ or even sooner at $F^2R_mF^2$. Change the pattern to obtain a tri-cycle of edges. [This tri-cycle is generally called the Varga–Fried tri-cycle after its discoverers Tamás Varga and Kate Fried — DS.]

2. Transform the tri-cycle of edges to one in which you can see each of the three moving edges in the middle layer between left and right. (In other words: the *unseen* edge in this layer should be the fixed one.) The solution is shown in Figure 2.42.

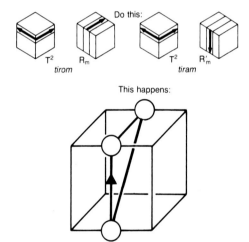

Figure 2.42   Cycling three edges in a middle layer

3. Our simple edge tri-cycle affects three edges of a middle layer (mathematically speaking: three parallel edges). Try to reduce the tri-cycles of edges in different relative positions to this simple case. When does this require one move only?

4. Explore other possibilities of reducing edge tri-cycles to the simple case!

2.8 Positioning edges by V-cycles

The tri-cycle called the V-cycle has two legs of the V starting at the top left and top right edges and meeting at the front down edge, as at the top of Figure 2.43. It can be reduced to the previous simple tri-cycle edge by a T (*to*) or a T′ (*ta*) turn (Figure 2.43).

Figure 2.43 shows that reducing the V-cycle to the simple situation of Figure 2.42, by T or by T′ produces two different cycles of the same three edges: an anticlockwise one and a clockwise one, as seen from the main corner.

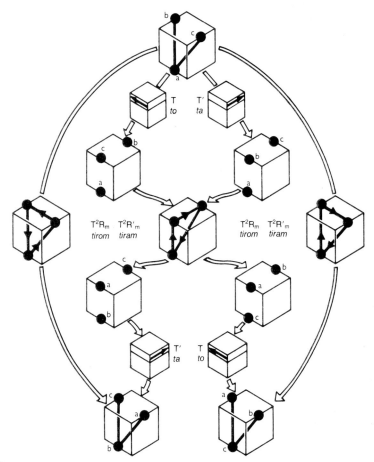

Figure 2.43 V-cycle transformed to three edges in a middle layer, anticlockwise or clockwise

Conjugation by *one* move normally lengthens a process by *two* moves, since it has to be undone at the end. But in this case T and T^2 combined give T' at the beginning and T' and T^2 combined give T at the end. In both cases the Figure 2.43 process is only longer than the original Figure 2.42 process in that one half turn (double move) is replaced by two quarter turns (single moves). Moreover the new processes have an elegant symmetrical structure with a half turn in the middle, flanked by two middle-layer moves, first up, then down, as seen from the front; and, at the two extremes, two halves of the central move. These latter moves are both clockwise moves if the cycle is intended to be clockwise, and are both anticlockwise if the cycle is to be anticlockwise.

Note that in Figures 2.44 and 2.45, in the proper position of the letter V both the horizontal and the vertical facelets keep their horizontality and verticality respectively. (We might say the edges keep this orientation even when moved.)

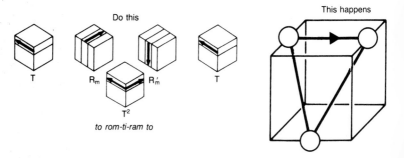

Figure 2.44   Clockwise V-cycle

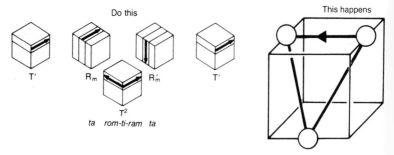

Figure 2.45   Anticlockwise V-cycle

Having started with the situation of three parallel edges (three edge cubes in one middle layer), we produced the V situation by only one move. Let us now go on and see how the cycling of other edge triplets can be similarly reduced. And, incidentally what are the possible relative positions of three edges?

Figure 2.46 provides the answer; although not all the possible transitions between edge triplets are indicated, those which are can be used to reduce all other triplets to the V situation in the shortest

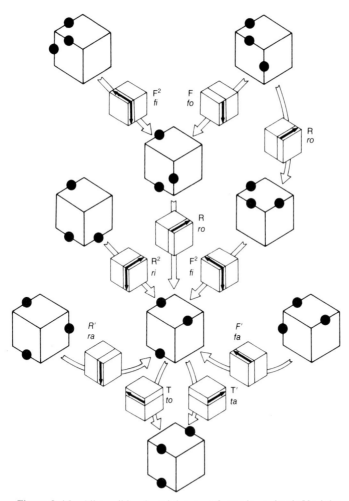

Figure 2.46   All possible edge triplets transformed to a simple V-triplet

possible way. (A diagram like this is useful, but it can not replace the familiarity gained by actually doing the turns.)

Using V-cycles, or other edge tri-cycles reduced to V-cycles, you can position all the edges. The following strategy is useful. Pick out an edge not yet positioned, such as *e* in Figure 2.47. Find its cubicle as defined by the centres. Four relative positions are possible: those marked by *a* are nearest to *e*, the one marked by *d* is the farthest. If

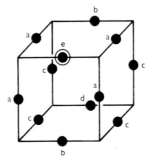

Figure 2.47 The four possible positions, (a), (b), (c) and (d), of edges relative to the fixed edge (e)

the cubicle of edge *e* is in a position *b* then both *e, b, u* and *e, b, v* form a letter V (Figure 2.48(a)). If it is in position *c*, then again there are two ways of obtaining a V triplet, by choosing either *x* or *y* (Figure 2.48(b)). If the cubicle is related to the edge as *a* or *d* is to *e*, then you will need to turn a face for it to fall into one of the two former situations; this move will have to be undone at the end.

The freedom of choice between the two possibilities (*u* and *v* or *x* and *y*) of completing the letter V may be used to avoid moving an edge already positioned. Or it can be used to position more than

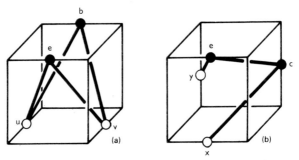

Figure 2.48 Two different pairs of edges form triplets in two different ways

one edge at a time. Three at a time is extremely rare, except that this always happens when every corner except three is positioned.

Let us now look at the two edge triplets at the top of Figure 2.46, which require two moves to be reduced to the V situation. In one, the three edges are adjacent to one corner. After a half-turn of any of the faces meeting at this corner it is easy to find another move which will form the letter V. In the top left position of Figure 2.46, F^2R (*firo*) are the two moves, to be undone by $R'F^2$ (*rafi*).

The other triplet of edges forms a chain, not on one face. There are several options. A half-turn of a face containing only one of the edges will always serve as the first move. Figure 2.46 suggests two other possibilities. In the position of the figure they are FR (*foro*) to be undone by $R'F'$ (*rafa*), and RF^2 (*rofi*) to be undone by F^2R (*fira*).

With the cube in your hand such tricks will happen quite naturally. Verbal descriptions are much clumsier and sound too intellectual. Children can usually find their way around a problem more easily by actually doing it.

Problems

1. An easy process, $(FR_m)^4$ or four *forom*s, flips four edges of a cube and does nothing else (exercise 2d of section 2.6). Transform the process so that three of the four edges are in the horizontal middle layer (i.e. in the layer below the top) and the fourth is either in the front or to the right.

2. Combine the ways of flipping four edges to produce a way of flipping two edges.

3. Explore the relative positions of four edges. Find ways of reducing them to the easy situations solved by $(FR_m)^4$, four *forom*s. How many moves are required in each case?

2.9 Flipping edges

The problems in the last few sections have developed a simple process for flipping four edges in a particular pattern, for example by $(FR'_m)^4$, four *foram*s. Flipping four edges in other relative positions will require preliminary moves. First, we study our particular pattern in other positions. We have three of the four edges situated in a middle layer, and the fourth edge is adjacent to only one of these three. To put it another way, the fourth edge to be flipped could be moved into the fourth place of the same middle layer by a single face turn (Figures 2.49 and 2.50 illustrate $(FT'_m)^4$, four *fotam*s).

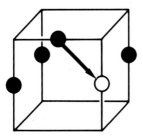

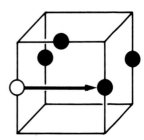

Figure 2.49 The fourth edge is moved into the horizontal middle layer

Figure 2.50 Restoring the previous situation

This second wording suggests the first move of the process — *moving the fourth edge into the fourth place of the middle layer*. Whatever the spatial location of the quadruplet of edges, if they resemble the above patterns, then this is our first move.

How about the second move? Well, this is simply a move which re-establishes the situation as it was before the first move. This does not mean undoing the first move. The situation to be re-established is this: three edges to be flipped in a middle layer with one edge adjacent to one of them. The first move filled the empty place in the middle layer with an edge to be flipped and at the same time it emptied another place in the same middle layer, for the example, the empty circle in Figure 2.50. Moving this not-to-be-flipped edge to where the empty place was before will in fact re-establish the previous pattern of the quadruplet of edges to be flipped.

Strangely enough, doing these two moves four times will flip the four edges, and we have an easily remembered general method for doing this.

Other quadruplets can be transformed into this particular pattern or its mirror image. Figure 2.51 may help — though practice with the cube is certainly much more helpful — to find the proper conjugation moves. Until you feel secure enough in reducing other quadruplets to the simple case, you should practise flipping edges in pairs instead of in fours. You should do this anyway if the number of edges to be flipped is not a multiple of four. Two situations of

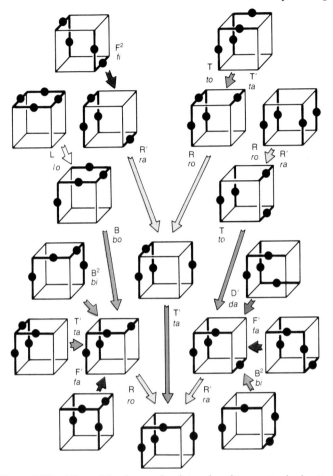

Figure 2.51 All possible edge quadruplets reduced to our standard pattern

flipping two edges are shown in Figures 2.52 and 2.53. The solutions combine two uses of problem 1 of the previous section ust as we combined results in Figures 2.20 to 2.22. The other two situations, in which the two edges are not on one face, can each be reduced to these two by one move. Flipping the edges is made easier by the fact that you apply the process to a nearly restored cube. Thus, even if you cannot exactly remember the preliminary conjugation moves you did, it should not be too difficult to undo hem at the end. Imagine a cube three moves away from the start.

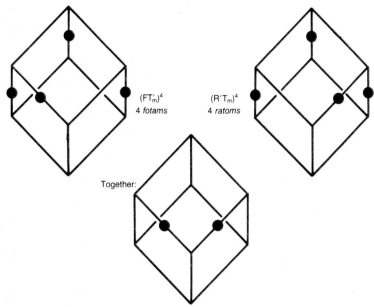

Figure 2.52   Flipping the two top edges at the main corner

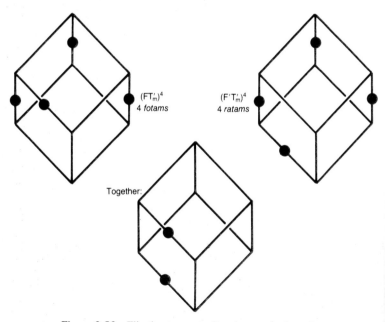

Figure 2.53   Flipping two opposite edges on the front face

Do you find it difficult to reverse these moves? If you do, then you need some practice; this is an interesting exercise in itself.

2.10 Summary: solving the cube calmly

See Figure 2.54.

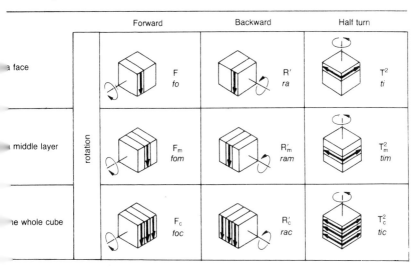

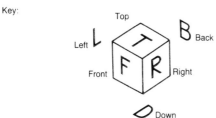

Figure 2.54 Notations: a summary

Positioning the corners. Swaps (Figure 2.55): a corner not in place should be exchanged with the one in its cubicle. Repeat this process. Each time the number of corners in place will be increased by one or by two. Tri-cycles speed up the procedure (Figure 2.56). These relative positions of two or of three corners can be produced by preliminary moves (to be undone at the end), if necessary.

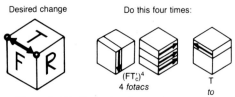

Figure 2.55   Swapping the top front corners

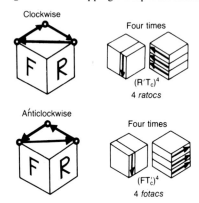

Figure 2.56   Cycling three top corners clockwise and anticlockwise

Orienting the corners. See Figures 2.57 and 2.58.

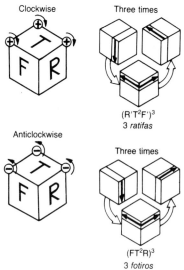

Figure 2.57   Twisting three top corners clockwise and anticlockwise

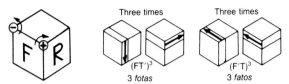

Figure 2.58 Twisting the two top front corners

Positioning the edges. V-cycles can be used. Creating a V-triplet may require a preliminary move. Then a wrongly placed edge should be moved into its cubicle while the edge in its cubicle should replace another wrongly placed edge. This procedure should be repeated; see Figures 2.59 and 2.60.

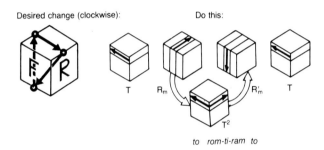

Figure 2.59 Cycling three edges clockwise

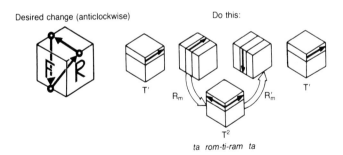

Figure 2.60 Cycling three edges anticlockwise

Orienting the edges. Figure 2.61 shows flipping four edges; Figure 2.62 shows flipping two edges.

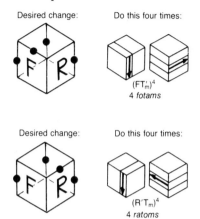

Figure 2.61   Flipping four edges in two congruent positions, mirror images of each other

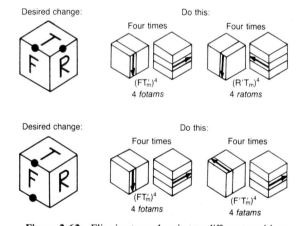

Figure 2.62   Flipping two edges in two different positions

2.11   An example followed through

We shall start with a thoroughly scrambled cube. The reader will be shown how to do the same scrambling and thus follow the procedure of unscrambling, too. Throwing a dice twice will be enough to define a move. We shall use the first number to tell us which face to move:

$$1\ 2\ 3\ 4\ 5\ 6$$
$$\text{F B R L T D}$$

and the second to tell us the type of move:

1 or 2	3 or 4	5 or 6
clockwise	anticlockwise	half turn

Moving middle layers is not included; this is just the same as a particular pair of face moves and a move of the whole cube. Moving the same face twice in a row will be excluded; a throw producing this will be ignored.

The following throws define 24 valid moves; the letters below each pair of throws indicate which they are. Crossed-out digits are throws which have been ignored.

46	62	56	22		34	~~3~~11	36	15		22	46	21	55
L^2	D	T^2	B		R′	F	R^2	F^2		B	L^2	B	T^2
li	*do*	*ti*	*bo*		*ra*	*fo*	*ri*	*fi*		*bo*	*li*	*bo*	*ti*

41	62	12	61		31	~~3~~54	16	41		14	24	42	~~4~~23
L	D	F	D		R	T′	F^2	L		F′	B′	L	B′
lo	*do*	*fo*	*do*		*ro*	*ta*	*fi*	*lo*		*fa*	*ba*	*lo*	*ba*

This number of turns will thoroughly scramble the cube.

If you start with a standard orientation—blue on top, red at the front—then this sequence of moves will produce the following pattern (Figure 2.63). (Take care; one false move is enough to spoil it. If your cube is coloured differently from the above, then you can still follow the example by preparing a code sheet for the colours.) After you have carried out a process, continue to hold the cube in the position reached unless otherwise stated. Care will be taken to perform the processes in the most usual spatial positions.

To position the corners the procedure described in section 2.6— but not included in the above summary—will be followed: forming

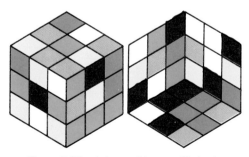

Figure 2.63 A thoroughly scrambled cube

a closed chain of blue corners (that is, gathering them into one layer) and then moving the blue centre into the same layer. Blue is, in fact, a good choice, since the blue corners form an open chain, only one move away from the closed chain. Holding the cube with the yellow centre on top and the orange in front, F (*fo*) closes the chain in the top layer. F'_m (*fam*) moves the blue centre into this same layer. This is what you see (Figure 2.64): (By the way, FF'_m is the same as F^2B' coupled with moving the cube itself by F'_c.)

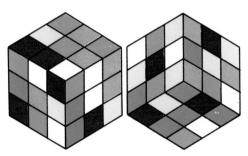

Figure 2.64   The blue corners forming a chain around the blue centre

The blue corners are all in the blue-centred layer, but none is in its own place. After T (*to*) not only the main corner (this is now the orange–blue–white corner) but also the corner opposite it (red–blue–yellow) will be positioned. The other two, opposite to each other, need to be exchanged. After R' (*ra*) this can be done by $(FT'_c)^4T$ or four *fotacs to*: then R (*ro*) undoes R' (*ra*). The blue corners are in position. You can see this in Figure 2.65.

Turn the cube upside down. The green corners are then in the top layer around the top centre, with only one of them in its own

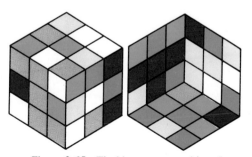

Figure 2.65   The blue corners positioned

cubicle, the white–green–orange corner. Rotate the cube about its vertical axis to make this the main corner. A clockwise cycle using $(R'T_c)^4$ or four *ratocs* now positions three corners (Figure 2.66).

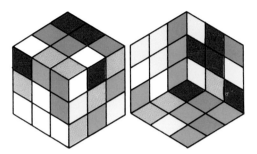

Figure 2.66 All corners positioned

One of the corners in the top layer, the orange–green–yellow one, is oriented correctly. Rotate the cube to make this the main corner. The other three can be correctly oriented by twisting them clockwise $(R'T^2F')^3$, three *ratifas* (see Figure 2.67).

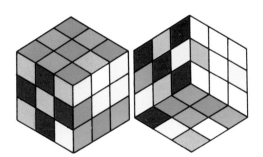

Figure 2.67 Six corners are correct

Two more corners need to be twisted, the blue–yellow–orange and the blue–yellow–red ones. Their two blue facelets should be moved from the yellow-centred to the blue-centred face. Keeping the cube with the yellow centre in front and the blue centre on top, this will be accomplished by $(FT')^3$ $(F'T)^3$, three *fotas*, three *fatos*. The corners are now all correct and the cube should look like Figure 2.68.

The number of edges (or of corners) which we would expect to be correctly positioned in a scrambled cube is one. In our cube, just

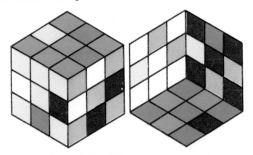

Figure 2.68   All corners are correct

one of the edges is in place, the blue–yellow edge and it is also oriented correctly. The green–yellow edge should now be moved to where the green–red edge is. In the cubicle of the green–red edge there is the blue–orange edge. These three edges do not yet form a V configuration, but one move is enough to turn them into a V. Holding the cube with the red centre on the top and the yellow in front, F′ or *fa* does the trick and $TR_mT^2R'_mT$ or *to romtiram to* provides the required forward rotation. F or *fo* repositions the corners which were moved away. Two edges, the green–yellow and the green–red ones, are now in their cubicles (Figure 2.69).

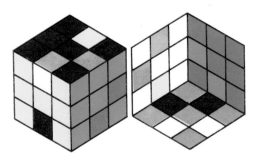

Figure 2.69   Three edges positioned

In the middle layer between left and right the opposite edges need to be swapped. This is done by $(F_m^2R_m)^2$ or two *fimrom*s (see section 2.6, problem 3(d)). This makes four edges go home, two oriented, the other two flipped; the flipped ones will be corrected later on.

The blue–orange edge should go where the green–white is and green–white's place is occupied by the blue–white edge. So we hav a V-triplet. Holding the cube with the blue centre on top and the white in front, an anticlockwise rotation, $T'R_m T^2 R'_m T'$ or *ta romtiram ta*, will position two of the three edges, giving us the position shown in Figure 2.70.

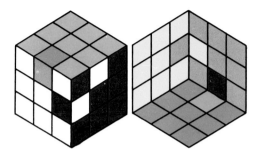

Figure 2.70   Nine edges positioned

The three edges not yet home can now be positioned simultaneously by one V-cycle. R^2 or *ri* creates an upside down V, F_c^2 (*fic*) turns it the right way up another $T'R_m T^2 R'_m T'$ (*ta romtiram ta*) positions the edges and L^2 (*li*) finishes the conjugation.

Flipping two edges is all that remains to be done. A half turn gets them into the same layer; so let us have them in the front layer, in the top and down positions. Then $(FT'_m)^4$ $(F'T'_m)^4$, or four *fotams* followed by four *fatams*, will flip them. A final half turn completes the solution of the cube.

2.12   Could it be shorter?

In this problem, we made 24 moves to scramble the cube. We used 79 single (or quarter) turns and 15 double (or half) turns to restore it, making 94 turns altogether $(14+21+33+26$, in the four stages). Counting a middle-layer move as one turn rather than two, the number is reduced to 78, or $14+21+25+18$. Yet ten random moves could easily have produced a pattern which would require more than one hundred turns to restore the cube, using our strategy.

Everybody dreams of a strategy which will restore in ten moves or less a cube which has been scrambled in ten. Why not? It is even possible (though not very likely), that the pattern reached by ten random moves could have been reached by some other process in fewer than ten moves, and then that way back would require fewer than ten. A perfect strategy should enable you to find the *shortest possible way from (and to) any state.*

Let us suppose that by means of a fabulous supercomputer, or some ingenious theory, or a combination of the two, it became possible to find the shortest way home from any pattern, the questions that would arise would include these.

How many patterns require the greatest number of moves?
Nobody knows. Maybe just one, maybe millions. [Evidence from smaller problems indicates a large number—DS.]

What is the greatest number of moves required?
Nobody knows the answer to this question either. Yet we do know one thing: most patterns require at least 18 moves. (How do we know this? Clues will be given in section 4.2.) Experts believe that even in the worst cases—the patterns furthest away from start—it should be possible to restore the cube in 20-odd moves, maybe 25, not more.

Our hypothetical method of finding the shortest way home is called God's Algorithm. It is perfect, its only disadvantage being that it has not yet been discovered!

The strategy of M.B. Thistlethwaite has the shortest known length. In the worst cases it requires 52 moves, approximately double the number of moves of God's Algorithm. [Work by students of D.E. Knuth has reduced this to 50—DS.] Thistlethwaite's strategy requires voluminous tables.

However, champion cubists do not have time to consult tables. They try to minimize the time for making both the moves and the decisions. Achieving a balance between the two, they probably do more than 100 moves in the most difficult cases to restore the cube. Those who just use some basic strategy may go well beyond that number, even double, not counting mistakes which of course may double the number yet again.

Though the strategy developed so far does not seem particularly long as far as the number of moves is concerned, there is nevertheless plenty of room for improvement.

The next chapter will enable the reader to improve his skill considerably. For the present let us concentrate on making the most of certain situations. The basic assumption in developing our strategy so far has been that the reader is dealing with a thoroughly scrambled cube. One can therefore assume, when restoring the corners, that the edges can be neglected, as their turn will come later. Or, by the same token, that while positioning the corners, you need not bother about their orientation. That too will be dealt with later. But what if, by some coincidence, many of the edges are all right, and the corners are not? Or many of the corners are all right although some are not positioned? Or everything is all right except the centres? (This actually means that nothing is all right, *except* the centres!) Would you still use all the same processes in the same strategy, from positioning the corners to orienting the edges? The following suggestions are worth considering.

(a) *Tri-cycling corners, keeping all the rest fixed.* Can we exchange just two corners? We have already done that, but ignoring the edges, which is what made it possible. When we had to keep the corners fixed, we found we could move three (or more) edges at a time, but never just two. Similarly, if the edges are to be kept fixed, then three is the minimum number of corners that can be moved. We can do this, leaving all the rest fixed in eight quarter turns. To make it easier to see, the processes in Figures 2.71 and 2.72 rotate

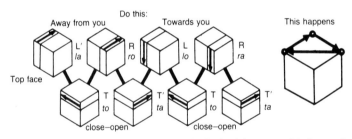

Figure 2.71 Cycling three top corners clockwise in eight turns, with the rest fixed

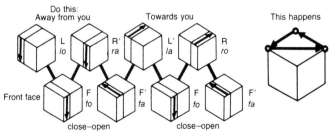

Figure 2.72 Cycling three top corners anticlockwise in eight turns, with the rest fixed

the same three corners as our old processes, $(R'T_c)^4$, or four *ratoc*s, clockwise, and $(FT'_c)^4$, or four *fotac*s, anticlockwise.

Clockwise: L'TRT' · LTR'T'
                 latorota lotorata
Anticlockwise: LFR'F' · L'FRF'
                 loforafa laforofa

The second process is the same as the former with its first and second half being exchanged and F replacing T. (That is, it is the same as LTR'T' · L'TRT', or *lotorata latorota*, conjugated by R'_c or *rac*, which means tilting the cube towards us—and, at the end, away from us.)

One interesting thing about these two tri-cycles is that they do not undo each other. If the second follows the first, then they twist two adjacent corners, the top front ones. If the order is reversed then they twist two opposite corners in the top layer, the neighbours of the main corner. (Try and see which way these corners get twisted.)

(b) *Twisting two corners.* We saw just now that this can be achieved in 8+8 turns without changing anything else. But 12 turns are enough if the corners are opposite each other on the same face. Moreover, this sequence of moves consists of two identical halves (Figure 2.73). [This process is due to E. Rubik—DS.]

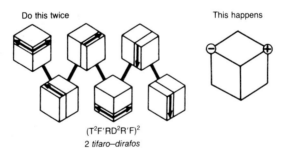

$(T^2F'RD^2R'F)^2$
2 *tifaro–dirafos*

Figure 2.73   Twisting two corners, with the rest fixed

How about two adjacent corners? Or two corners furthest apart, at the ends of a space diagonal? Conjugates of the above will do: one turn moves the two corners to make them opposite, which must be undone after the twists. This seems to require 14 turns in all, but it can be reduced to 13 for adjacent corners and 12 for corners furthest apart (see problem 2).

(c) *Cycling the centres.* In solving problem 3f in section 2.6, we found a pattern in which each face has one colour, except for differently coloured centres (the pattern looks like a letter O or a dot on each face). Our strategy so far is fast for restoring the corners of this pattern but not the edges. So it would be nice to be able to move the centres, with all the rest fixed.

Reversing the process found in problem 3f is a method for doing this, but the following is better, and it is independent of the spatial position of the cube. Move a centre of any colour into the middle of the letter O of the same colour. This transforms the O into an H on a background of a different colour.

Move the centre with the second colour into the middle of the H. A letter I appears, as in the third cube of Figure 2.74.

Move the letter I between two stripes of the same colour. This face and the opposite one are now correct.

A final middle layer move restores everything else.

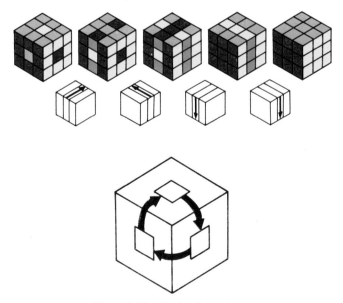

Figure 2.74 Moving the centres

Two of the three middle layers were moved alternately by two opposite quarter turns. Figure 2.74 shows the moves and the result. Since opposite centres always remain opposite, they are also moved in a corresponding way.

Problems

In 1 and 2, keep everything fixed which is not asked for.

1. You want to twist the same two corners that are twisted by $(T^2F'R \cdot D^2R'F)^2$ or two *tifaro-dirafo*s, but both of them in the opposite sense. What should you do?

2. (a) Twist two adjacent corners in 13 turns.

(b) Twist two corners furthest apart in 12 turns.

3. Do the following and see what it is useful for: $(RT_c)^4T \cdot (F'T'_c)^4T'$ (four *rotoc*s *to*, four *fatac*s *ta*).

4. Find a process which performs the same tri-cycle as the above, but the other way around.

5. You see dots (or the letter O) on four faces, not six. How do you restore the cube?

2.13   Transforming formulae

Suppose you know a process for working on, say, the left layer but you would prefer to work on, say, the top layer. How do you transform the process? How do you find the mirror image of a process, that is, reflect it in a given plane of symmetry?

We have met this type of problem before. What this section does is to survey and complete our knowledge of transforming formulae.

Symmetries exchanging the front and the right-hand layers include:

this	becomes	this
$(RT_c)^4T \cdot (F'T'_c)^4T'$		$(F'T'_c)^4T' \cdot (RT_c)^4T$
(four *rotoc*s *to*, four *fatac*s *ta*)		(four *fatac*s *ta*, four *rotoc*s *to*)
$(FT'_c)^4$		$(R'T_c)^4$
(four *fotac*s)		(four *ratoc*s)

The latter can also be expressed in face turns only:

FLBR	R'B'L'F'
(*foloboro*)	(*rabalafa*)

These are both inverse and mirror images of each other—the mirror placed as in Figure 2.75. The 'glossary' of the reflection is shown in Figure 2.76. In formulae this reflection exchanges F for R and L for B. Further, the sense of rotation is reversed. This means dropping the prime (′) where there is one and putting one where there is none. In the pronunciation the changes are:

$$f \leftrightarrow r \qquad\qquad l \leftrightarrow b \qquad\qquad o \leftrightarrow a$$

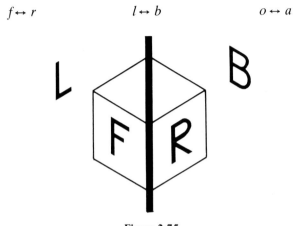

Figure 2.75

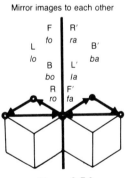

Figure 2.76

T and D remain unaltered in this kind of reflection, but their senses are reversed. As an example, FLBR · T or *foloboro to* (which swaps two corners) becomes R′B′L′F′T′ or *rabalafa ta*. This is how $(FT'_m)^4$ or four *fotam*s is transformed into $(R'T_m)^4$ or four *ratom*s: T remains T, only F and R change roles as also the primes (or *a*s) with the lack of primes (or *o*s).

Sometimes it is necessary to transform a formula to make it easier to see what is going on when making the moves. As an example, let us look for a process flipping four visible edges. Figure 2.77 shows that the cube can be held so that at least one facelet of each of the four edges in our particular configuration can be seen at the same time.

Figure 2.77   Flipping four visible edges

In the situation of $(FT'_m)^4$, four *fotam*s (see Figure 2.61, section 2.10), the layer containing three of the four edges was horizontal, now it is vertical. Previously, the fourth edge was on top of that layer. Now it is to the right. Clearly T must become R. What does F become? F was the only face with two of the four edges adjacent. In Figure 2.77, this is face D. This completely determines the transformation of faces.

The sense of rotation remains to be determined. The fourth edge, not in the same middle layer as the other three, is moved into that middle layer by F or *fo* and by D or *do* in the two situations. This means that there is no change in the sense of rotation, no need for a reflection; the whole cube can be moved so that the four edges to be flipped are in the position shown in Figure 2.77.

(If you were able to see that before this lengthy explanation then you have excellent spatial vision.) The process of flipping the four edges marked in Figure 2.77 will be $(DR'_m)^4$ or four *doram*s.

What is the effect of the inverse process $(R_m D')^4$ or four *romda*s?

Since the inverse of a flip is a flip, both processes have exactly the same effect.

Figure 2.78 and 2.79 summarize the transformation of formulae visually, both without and with symmetry. Exchanging *o* with *a* is the same as adding or taking away a prime (′) symbol. Such changes take place if reflection is involved; just moving the cube entails no such change. Whether or not the transformation comprises a

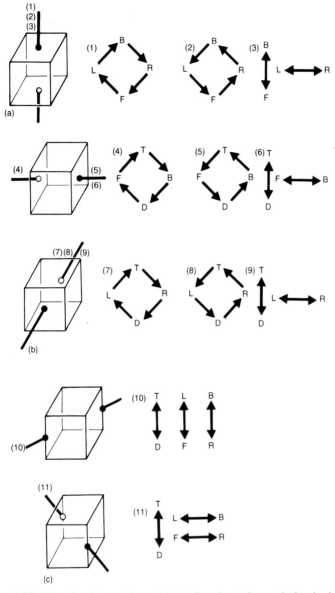

Figure 2.78 Transforming rotations: (a) rotating about the vertical axis through centre pieces (1–3); (b) rotating about horizontal axes through centre pieces (4–9); (c) rotating about diagonal axes through vertical edges (10–11); (d) rotating about the diagonal axes through horizontal edges (12–15); (e) rotating about the diagonal axes through corners (16–23)

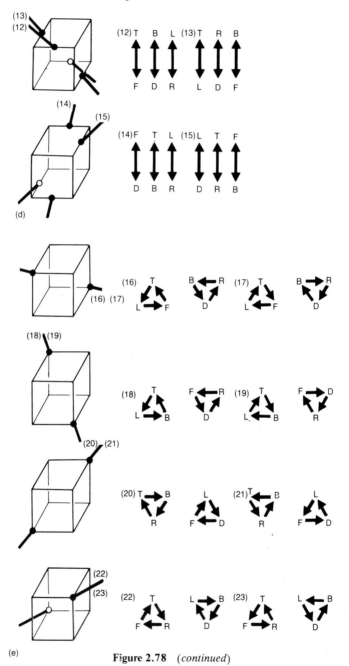

Figure 2.78   (*continued*)

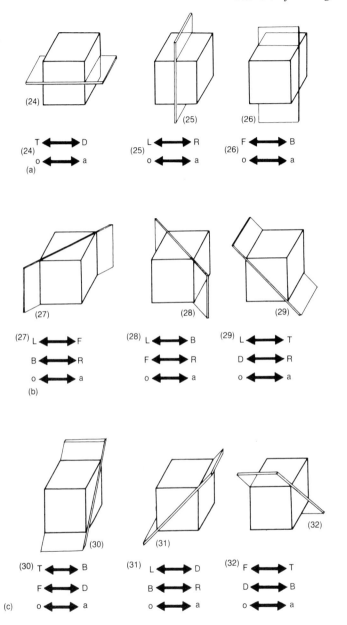

Figure 2.79 Transforming reflections: (a) horizontal/vertical planes through middle layers (24–26); (b) diagonal vertical planes through edges (27–28); (c) diagonally tilted planes through edges (29–32)

reflection, the power ^{2} (or letter i) is unchanged. The capital letters not mentioned in a transformation are also unchanged.

Diagrams (1), (2), (3) in Figure 2.78 correspond to T_c, T'_c, T^2_c. Diagrams (4)–(6) and (7)–(9) are the analogues for R_c and F_c. In diagrams (16)–(23), the even diagrams correspond to clockwise rotations as viewed from the labelled end of the axis.

The list of transformations without reflection is complete, and the list of pure reflections is complete. However, combining a spatial move with reflection generates some transformations which are not listed (Figure 2.79). One such is exchanging opposite faces and hence also the sense of rotation:

$$T \leftrightarrow D, \qquad L \leftrightarrow R, \qquad F \leftrightarrow B, \qquad o \leftrightarrow a.$$

2.14   The message of the cube

Suppose you have developed the skill of solving the cube within a minute. You can do it almost automatically. Now you can start doing something else: taking two differently randomized cubes, you can try to make one the same as the other, by turning only one of them; that is, you try to copy the other cube. This is a tricky exercise, but it is not impossible if you have a clear vision of the movements of the cubes.

Can you apply your skill in solving Rubik's cube to the $2 \times 2 \times 2$ cube, the $3 \times 3 \times 2$ Magic Domino (Figure 2.80), or to other similar toys? This shows whether your skill is narrow or wide, rigid or elastic. Take the domino (Figure 2.80). A little experimentation is enough to show which cube processes can and which cannot be applied to it. Those containing only double turns of two pairs of opposite faces can be applied; the others cannot. After some more experimentation you learn that a cube process consisting of double turns only, such as $(F^2 R^2)^3$ or three *firis*, corresponds to two different domino processes as in Figures 2.81 and 2.82.

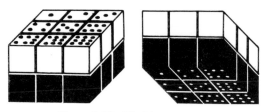

Figure 2.80   The Magic Domino

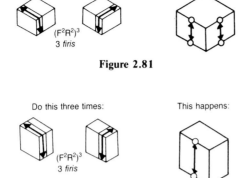

Do this three times:

$(F^2R^2)^3$

3 *firis*

This happens:

Figure 2.81

Do this three times:

$(F^2R^2)^3$

3 *firis*

This happens:

Figure 2.82 Here only one pair of edges is exchanged—the other pair is not present

This should be enough to indicate how your knowledge of the cube can be applied to the Magic Domino.

The $2 \times 2 \times 2$ cube is almost trivial. The lack of both edges and centres means you only have to deal with the corners. However, the fact that there are no centres may be quite disturbing for a while!

By applying coloured stickers in various ways you can make each of these toys present different problems, some of them easier, some more difficult. As an example, you can have no colour on the corners (or the edges [or the centres—DS]), just leaving them black. Or you can put the same colour on five faces and a second colour on the sixth. Or you can have the opposite faces the same colour. Whether or not you can use your knowledge in such cases is a test of how substantial it is.

Why is it worth acquiring cube skills? To be able to tackle other toys of the same family? No, there is more than that. Rubik's cube is a mathematical toy. Whether or not you like or are good at mathematics, it provides you with some mathematical background.

You learn group theory, for example, without even knowing what group theory is. (Strangely enough, it *is* possible to learn something without even knowing what it is.) A student of group theory may have learned its axioms, a lot of theorems, and even the proofs of the theorems, yet still remain totally unaffected by the ideas of this discipline. On the other hand, from the insight gained by dealing with the cube it is possible to grasp, with a little help, the main ideas of basic group theory.

Group theory is an algebraic discipline and, in this respect, the ideas generated by the cube are algebraic. In addition, many geometric concepts are involved—spatial symmetry, to mention but one. Number theory appears now and again. Think of the least common multiple which we met during our first discovery tour, or of congruences module 2 or 3 which determine the possible patterns of twists and flips. We also met combinatorial problems, probability and logic, and all these interdependently!

So you can see that working with the cube paves the way to mathematical thinking and is actually a mathematical activity in itself—or rather, mathematical and scientific—we shall discuss some of its connections with physics in Chapter 5.

What does the cube teach us? Beyond the answers 'playing with similar toys', 'mathematics', 'physics', something more general can be said. It teaches a particular way of thinking, useful in the theoretical fields mentioned above and in other fields. This includes collecting experience, systematizing it, drawing conclusions, examining the results, assessing their worth, searching for further tools and inventing notation systems, to name but a few.

Solutions

Solutions to section 2.2

0. The order of turning two opposite (parallel) layers can be reversed, but the effect is just the same. For instance, LR' or *lora* is the same as R'L or *ralo*. The inverse of either of them can be called RL' (*rola*) or L'R (*laro*).

1. After $(FR)^7$ or seven *foros*, the edges are all correct; hence also after 14, 21, 28, ... *foros*. The first time all the corners are correct is after $(FR)^{15}$ or 15 *foros*; then again after 30, 45, 60, ... *foros*. The cube will be solved when the two sequences meet, after $(FR)^{105}$ or 105 *foros* and again after 210, 315, ... *foros*.

2. As for problem 1.

3. Instead of the multiples of 7 and 15, we have the multiples of 7 and 9. The cube will first be solved after $(FR')^{63}$ or 63 *foras*, then after 126, 189, ... *foras*.

4. As for problem 3.

5. This time the corners are restored before the edges, at $(FR^2$

or six *foris*; then again at 12, 18, 24, ... *foris*. For the edges the numbers are 10, 20, 30, ... The cube will be restored first at $(FR^2)^{30}$ or 30 *foris*. These numbers are the same when repeating F^2R or *firo*.

6. As for problem 5.

7. Select a corner which is out of position. After F^2R or *firo* the main corner is also out of position. The red–blue–yellow (*rby*) corner should be in this cubicle, but we find the blue–red–white (*brw*) cube there instead. (Note that the facelets have been listed in the same order as the centres which define the cubicle; the order both times was front, top, right.) The *brw* cubicle is occupied by the *gry* cube. And so on:

rby, brw, gry, ogy, boy, rgw, (rby).

The tour is finished: the red–green–white cubicle is occupied by the red–blue–yellow cube, whose cubicle was our starting point. The parentheses around *rby* indicate that the list is complete without it. The order of the letters gives information about the orientation in which each corner occupies a cubicle. The fact that (*rby*) at the end contains the same letters in the same order indicates that having done F^2R or *firo* six times, that is, after six steps forward in this round trip, the red–blue–yellow corner will get back to its own cubicle in the same orientation as it was, as defined by the centres. The same is true for the other five corners partici-pating in the round trip. The remaining two corners are not affected and keep both their position and their orientation.

Of the 12 edges, five are left unchanged by F^2R, *firo*: *bw, bo, wo, wg, og*. Repeating F^2R the same thing happens again and again: these which were not changed the first time will remain unchanged.

Five of the remaining seven form a cycle:

by, rw, ry, gy, oy, (by),

while another, smaller cycle consists of these two:

rb, rg, (rb).

This smaller cycle means that these two simply change places. After $(F^2R)^5$ or five *firos* each edge in the longer cycle will be correct—in its own cubicle and in the correct orientation—but the two edges of the small chain will be swapped. This is why $(F^2R)^{10}$ or ten *firos* are needed for every edge to get back correctly. How the previous six and this ten yield 30 (the number of F^2R or *firo* turns necessary to get back to start) is now clear. So looking at the cube after just one F^2R or *firo* gives enough information to work out all that has been found out experimentally.

If the initial triplet (or pair), of colours first reappears in a cycle with a different orientation (as may well be the case), then you continue and find that a cycle three times (or twice) as long will re-establish the original orientation. In the future you will probably stop at the first reappearance and draw conclusions from the incomplete chain.

8. Both $(FR)^5$ or five *foro*s and $(F'R')^5$ or five *fara*s will re-position every corner, but they twist six of them, all in the same way. The direction is anticlockwise in the first process, clockwise in the second. Fewer turns are enough for the next two: both $(FR')^3$ or three *fora*s and $(F'R)^3$ or three *faro*s will reposition every corner with three corners twisted in one direction and three in the other. Both processes twist the same six corners but not in the same way. Repeating FR^2 (*fori*), F^2R' (*fira*) or $F'R^2$ (*fari*) leaves all the corners in their proper orientation when they return to their cubicles.

9(a). $(FTR)^4$ or four *fotoro*s does a clockwise twist to the main corner and its two adjacent ones in the top layer.

9(b). $(R'T'F')$ or four *ratafa*s twist the same three corners anti-clockwise.

9(c). $(FT^2R)^3$ or three *fotiro*s gives an anticlockwise twist to all the top corners except the main corner.

9(d). $(R'T^2F')^3$ or three *ratifa*s—the same corners are twisted clockwise.

Solutions to section 2.3

1. As suggested by the figure, of the two front top corners, the one to the left needs an anticlockwise twist and the one to the right needs a clockwise twist. For the process $(FT')^3$ $(F'T)^3$ or three *fota*s three *fato*s should be applied.

2. The problem is to twist the same two corners, but in the opposite directions to those in problem 1. One solution is to do the above process twice. Another is to do the inverse process, $(T'F)$ $(TF')^3$, three *tafo*s, three *tofa*s. A third is to use the original process but holding the cube differently, with the two corners changing places. (Move the front face to the top while the top becomes the front.)

3(a). See Figure 2.83. Twisting two corners requires 12 turns; so does twisting the other two. Rotating the cube in between is not counted.

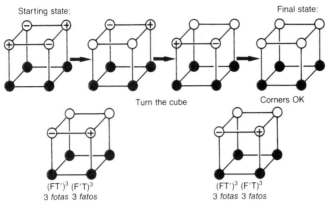

Figure 2.83 Twisting all four top corners (case a) in 24 turns

3(b). Doing $(FR')^3$ $(F'R)^3$ or three *foras*, three *faros* twice also solves this problem. The difference lies in moving the cube between the two processes.

Both (a) and (b) can be resolved in 9+9 turns instead of 12+12 by twisting three rather than two corners at a time as in Figures 2.84 and 2.85. See also problem 4.

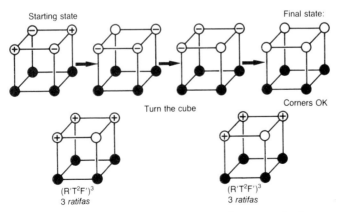

Figure 2.84 Twisting the same corners (case a) in 18 turns. In the first step, only one corner should be oriented

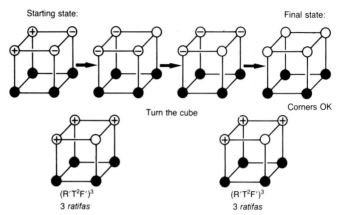

Figure 2.85   Twisting all four corners (case b) by twisting three corners at a time. In the first step, only one corner should be oriented

4. See Figure 2.86 and 2.87. These are short solutions to problems 3(a) and (b).

5. They twist the main corner and those adjacent to it in the top layer. Edges, too, are only moved in the top layer. The two processes are inverses.

6. This and problem 5(a) are symmetric left and right changing places both in what we do and, consequently, in the results. Front remains front, top remains top, hence the letters F and T keep the same positions in the formulae. The sense of rotation is reversed: *o* and *a* (no prime and prime) replace each other, and twists become opposite twists.

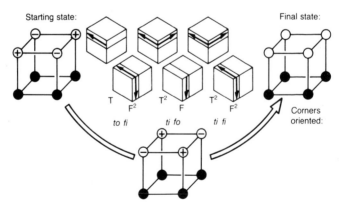

Figure 2.86   Twisting all four corners (case a) in six turns

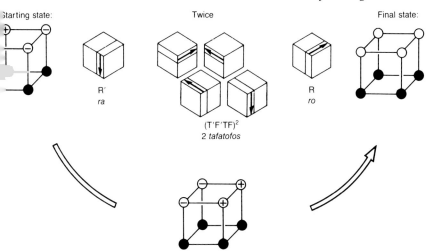

Figure 2.87 Twisting all four corners (case b) in ten turns

7. They are each the inverse of the other (one undoes the other) and are also mirror images. Instead of left and right, the faces changing place are: front and right, left and back. Rotations are also reversed. The plane of symmetry passes through the diagonals of the top and down faces.

Solutions to section 2.4

1. Here is a solution: $R^2(FT')^3 (F'T)^3R^2$, *ri*, three *fota*s, three *fato*s, *ri*.
2. Again, conjugation by R^2 (*ri*) works.
3. This can be reduced by B (*bo*) to the previous situation. Thus BR^2 or *bori* precedes the twist, which is then followed by its inverse, R^2B' or *riba*.
4(a). Start and end with T^2, *ti*.
4(b). Conjugate the process under (a) by R^2, *ri*. This means starting the process with R^2T^2, *riti* and ending it with T^2R^2, *tiri*.
4(c). As above but conjugated by F^2, *fi*.

Solutions to section 2.5

1. These can be reduced to swapping two adjacent corners in the same way that twisting two non-adjacent corners was reduced to

twisting two adjacent ones. In the situation illustrated in Figure 2.27 we started with R', *ra*, in Figure 2.30 with R^2, *ri* (see section 2.4, especially problem 1).

2. Again, reducing these to the situation of three corners in one layer is done in the same way as for the case of twists (see problems 2 and 3 in section 2.4).

3. Two tri-cycles are enough to reach 7 from 2: a cycle will position corners c, g, b, then e, f, and d, after an initial T', *ta*, subsequently reversed by T, *to*.

Solutions to section 2.6

1(a). If only middle layers are moved then the corners are not affected. Hence the four corner facelets on every face remain, as at start, the same colour.

1(b). Consider two edge facelets which are opposite on a face. Moving a middle layer keeps them opposite, though on another face. Hence opposite facelets on every face remain, as at start, the same colour.

1(c). One two, three, or four. The four different colours which can occur are those (i) of the centre, (ii) of the corners, and (iii) and (iv) of the two pairs of opposite edge facelets. Any two, any three and, of course, even all four can be the same.

1(d). See Figure 2.88.

2(a). The effect of $(F^2R_m^2)^2$ or two *firims* is almost the same as that of $(F^2R^2)^3$ or three *firis*: both of them exchange two top edges with the down edges below them. The only difference is that the present process affects two front and two back edges, instead of front and right. See Figure 2.89 for a graphic comparison.

2(b). $(FR_m^2)^6$ or six *forims* swaps both pairs of opposite corners in the front layer. An additional F^2 (*fi*) replaces the four corners, but swaps the opposite edges instead. (Figure 2.90).

2(c). $(F^2R_m)^4$ or four *firoms* causes an edge tri-cycle in the middle layer between left and right (Figure 2.91).

2(d). $(FR_m)^4$ or four *foroms* restores everything except four edges, which are just flipped (Figure 2.92).

All of these are useful in restoring the edges, especially the last two.

3(a). $F_m^2R_m^2$ or *fimrim* changes too much (it produces a nice pattern), yet $(F_m^2R_m^2)^2$ or two *fimrims* change nothing. This is the shortest round trip except for turning faces about one axis.

Key to the colours:

C	s_2	C
s_1	c	s_1
C	s_2	C

(1) Four colours on the face in this pattern

We have three colours if

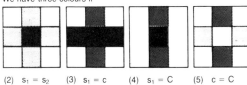

(2) $s_1 = s_2$ (3) $s_1 = c$ (4) $s_1 = C$ (5) $c = C$
                   or $s_2 = c$ or $s_2 = C$

We have two colours if

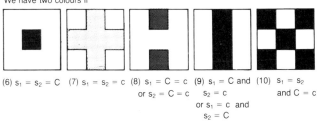

(6) $s_1 = s_2 = C$ (7) $s_1 = s_2 = c$ (8) $s_1 = C = c$ (9) $s_1 = C$ and   (10) $s_1 = s_2$
                          or $s_2 = C = c$ $s_2 = c$ and $C = c$
                                               or $s_1 = c$ and
                                                  $s_2 = C$

We have one colour if

$s_1 = s_2 = c = C$

Figure 2.88 Moving only the middle layers can produce 11 different face patterns; C denotes the four corner facelets; s_1, the left and right edges; s_2, the top and bottom edges; and c, the centre

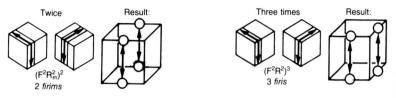

Figure 2.89   Solution to 2(a): closely related swaps of two pairs of edges

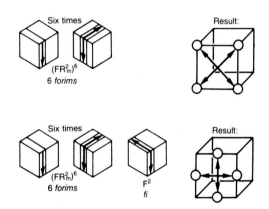

Figure 2.90   Solution to 2(b): swapping two pairs of corners transformed to swapping two pairs of edges

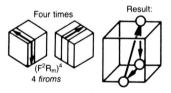

Figure 2.91   Solution to 2(c): a tri-cycle of edges (it is impossible to move fewer than three edges at once)

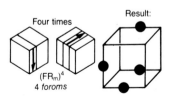

Figure 2.92   Solution to 2(d): the marked edges are flipped

3(b) and (c) give the same pattern, the *pons asinorum* [or 6X—DS]. Whatever the order of the three middle-layer half-turns, the result is invariably this (see section 4.2). Repeating these moves restores the cube.

3(d). $(F_m^2 R_m)^2$ or two *fimrom*s affect the edges of the same middle layer, the one between left and right, as does $(F^2 R_m)^4$, four *fimrom*s, but they move all four, exchanging the opposite ones.

3(e). $(F_m R_m^2)^2$ or two *fomrim*s do the same as 3(d), but in another middle layer, the one between front and back.

3(f). $(F_m R_m)^4$ or four *fomrom*s rotate the seen centres and the unseen ones, clockwise and anticlockwise respectively.

4(a). $F_m^2 T_c^2$ or *fimtic* is just a middle-layer move, undone as soon as it repeated.

4(b). $(F_m^2 T_c)^2$ or two *fimtoc*s does exactly the same as $F_m^2 R_m^2$, *fimrim*, in problem 3(a) except for a rotation of the whole cube.

4(c). $(F_m T_c)^4$ or four *fomtoc*s gives a six-dot pattern like $(F_m R_m)^4$ or four *fomrom*s. It leaves the cube in a different position so that the three colours in the centres are not the same as the three colours seen on the rest of the three faces around the main corner. In other words, it has rotated the centres about a different spatial diagonal of the cube.

Solutions to section 2.7

1. $F^2 R_m F^2$ or *firomfi* leaves the left and right layer intact, with changes only in the middle layer between the two—see Figure 2.93. R_m' or *ram* positions the centres and one of the four edges in this layer. So we have a process $F^2 R_m F^2 R_m'$ or *firom firam* producing an

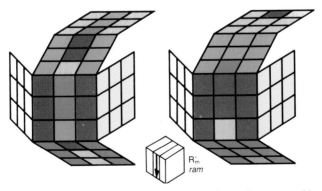

Figure 2.93 Turning a middle layer brings home almost everything

edge tri-cycle similar to $(F^2R_m)^4$, four *firom*s, but in half as many moves.

2. The edge of the middle layer between left and right not affected by $F^2R_mF^2R_m$ or *firom firam* is the top back one. This can be moved to the unseen position of this middle layer by R_c or *roc* which means tilting the cube away from us. This replaces front with top (F by T) but keeps right (R) in place, to give the effect of cycling the three visible edges of this middle layer. The process which does this is $T^2R_mT^2R'_m$ or *tirom tiram*.

3. Let us consider the pattern reached by applying $T^2R_mT^2R'_m$ or *tirom tiram* to start and make moves from it. A quarter turn of the front or top layer, or those parallel to them, changes the edge triplet which is affected, but any other single move keeps the relative position of the three edges unchanged. If their relative position changes, it always becomes the same new relative position: two of the three edges (in the mathematical sense) are parallel, the third is perpendicular but not adjacent, to them. This is the only relative position one move away from the simple one with three parallel edges (that is, with all the edge cubes in the same middle layer).

4. Each of the remaining edge triplets need two or (in two cases) three moves to reach the simple case with three parallel edges. The relative position suggested by the previous problem can be reached in one, or at most two, moves from any other edge triplet (for details see section 2.8).

Solutions to section 2.8

1. $(FR_m)^4$ or four *forom*s flip three edges in the layer between left and right. The one not flipped in this layer is the top front (*tf*) edge; the one flipped but not in this layer is the left front (*lf*) edge. Tilting the cube by F_c or *foc* makes the left and right faces horizontal. The left front edge becomes top front (*lf → tf*). F remains F but R_m becomes D_m which is the same as T'_m and the latter is standard. So the whole process becomes $(FT'_m)^4$, four *fotam*s.

Reflecting every move in the horizontal plane which bisects the cube, the resulting process flips the same three edges in one layer the fourth becoming the down front (*df*) instead of the top front (*tf* edge. Reflections change the sense of rotation, which would yield $(F'D_m)^4$ or four *fadom*s, but D_m is the same as T'_m, therefore the process becomes $(F'T'_m)^4$ or four *fatam*s. So far the fourth edge to be flipped (in addition to the three in the horizontal middle layer was in the front layer. A reflection exchanging front with right transforms the above processes to $(R'T_m)^4$ or four *ratom*s and $(RT_m)^4$ or four *rotom*s respectively. Figure 2.94 shows each of these three variants.

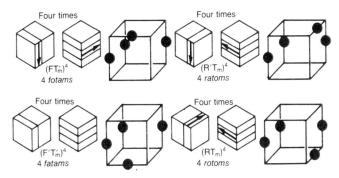

Figure 2.94 The marked edges are flipped

2. $(FT'_m)^4$ $(R'T_m)^4$ or four *fotam*s, four *ratom*s flips two adjacent edges in the top layer, the ones adjacent to the main corner. The point is that these are the only edges which are flipped once by the two combined processes. Other edges are flipped twice which amounts to no flip at all. Similarly, $(FT'_m)^4$ $(F'T'_m)^4$ or four *fotam*s, four *fatam*s, flip two opposite edges in the front layer (see Figures 2.52 and 2.53 in section 2.9).

3. If configurations equivalent to each other by some symmetry are not distinguished, then there are 18 different relative situations of four edges of a cube. Figure 2.51 in section 2.9 illustrates them and shows how to reduce the others to a simple situation in one, two or three moves. (This can be done in many other ways. The diagram is by no means exhaustive.) The simple situation is the one restored by the process $(FT'_m)^4$, four *fotam*s; see problem 1.

Solutions to section 2.12

1. One solution is repeating the process $(T^2F'R \cdot D^2R'F)^2$, two *tifaro-dirafo*s, that is doing four *tifaro-dirafo*s. Another is doing the inverse process once: $(F'RD^2 \cdot R'FT^2)^2$, two *farodi-rafoti*s. Last but not least, you may simply exchange the two corners by T_c^2 or *tic*, and then do $(T^2F'R \cdot D^2R'F)^2$, two *tifaro-dirafo*s.

2(a). If the main corner is the one to be twisted anticlockwise and

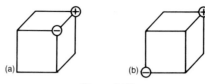

Figure 2.95

the other top right corner needs a clockwise twist as in Figure 2.95(a), then $F'(T^2F'R \cdot D^2R'F)^2F$ solves the problem. FF at the end merge to become one F^2 or *fi* move, so there are only 13 turns.

2(b). In the position of Figure 2.95(b) start the process with F to make the two corners opposite in the top layer. Instead of undoing F by F', omit the last F as well as F', so there are only 12 turns.

3. The three edges adjacent to the main corner are rotated clockwise. The process takes ten quarter turns, not counting moving the cube. This is less than reducing the tri-cycle in two moves—to be undone at the end—to either $TR_mT^2R'_mT$ (*to rom-ti-ram to*) or $T'R_mT^2R'_mT'$ (*ta rom-ti-ram ta*).

4. Inverting, you get $T(T_cF)^4 \cdot T'(T'_cR')^4$ or *to* four *tocfo*s, *ta* four *tacra*s. By reflection: $(F'T'_c)^4T' \cdot (RT_c)^4T$, four *fatac*s *ta*, four *rotoc*s *to*. This is easier: the two halves of the process change places.

5. A solution can be derived from the process of exchanging two pairs of opposite edges in a middle layer. A half turn of this middle layer positions the four edges, thus exchanging two pairs of opposite centres instead. This process was met in problem 3b of section 2.6 as $(F_m^2R_m)^2$, two *fimrom*s; this should be carried out on the left–right layer. Completing it, as suggested, by R_m^2, transforms it to $F_m^2R_mF_m^2R'_m$ or *fimrom-fimram*. Again, as for the six-dot pattern, the process can be described independently of spatial positions. The first move of restoring the cube from the six-dot pattern was to bring one of the centres to where it belongs. This required a quarter turn of a middle layer. In the case of the four-dot pattern the first move will again be to bring a centre to where it belongs, but now this requires a half turn of a middle layer—not the middle layer which contains the four dots. Again, a letter H appears. The second move is to destroy this letter by a quarter turn of a middle layer. The third move is the same as the first. The fourth will be evident; it is the opposite move from the second.

3

RESTORATION METHODS AND TABLES OF PROCESSES

Gerzson Kéri

3.1 Notation

The basic $90°$ moves or quarter turns of faces are combined into different simple and complex moves, which are called processes or sequences. This will be denoted by sequences of capital letters with or without indices. As indices we may use superscripts, subscripts or both. We use the capital letters F for front, B for back, R for right, L for left, T for top, and D for down. In general, a capital letter, without subscripts denotes a move of a single layer, but with subscripts, a move of two or three layers.

(a) A $90°$ clockwise turn of any face will be denoted by the corresponding capital letter without indices: F, B, R, L, T, D (clockwise is always defined as one looks directly at the face).

(b) A $90°$ anticlockwise (counterclockwise) turn of any face will be denoted by the corresponding capital letter with a prime superscript: F', B', R', L', T', D'.

(c) A $180°$ turn of any face will be marked by the corresponding capital letter with a superscript 2: $F^2, B^2, R^2, L^2, T^2, D^2$.

(d) We shall use a special notation for moves of two parallel faces, such as FB'. These movements will be called slice turns and denoted by a subscript s after the corresponding capital letter: $F_s=FB'$, $F_s'=F'B$, $F_s^2=F^2B^2$, $R_s=RL'$, $R_s'=R'L$, $R_s^2=R^2L^2$, $T_s=TD'$, $T_s'=T'D$, $T_s^2=T^2D^2$. (Note that $F_s=FB'$ turns the F and B faces as a unit.)

(e) To denote the move of a middle layer we shall use a subscript m after the corresponding capital letter. Thus F_m means moving the

95

layer between F and B in the same direction as F moves the front face. F'_m denotes the move of that layer in the opposite direction, and F^2_m denotes the 180° turn of this layer. R_m, R'_m, R^2_m, T_m, T'_m and T^2_m are defined similarly. A quick way of performing middle layer moves is the following. To perform F_m, first make a F' move, then turn back face F *and* the middle layer together. (Alternatively, turn F and the middle layer clockwise together and then do F'.) To perform F'_m, first do an F move, then turn back face F *and* the middle layer simultaneously.

If we are not concerned with the spatial position of the cube as a whole, then the move of a middle layer can be achieved by slice turns. In this sense F_m is the same as F'_s. The combination of F_m and F_s, however results in a spatial move, a tilting of the whole cube, while the positions of the cubes relative to one another remain unchanged. The main difference is, that slice turns do not change the positions of the centres, while moving a middle layer does. Slice turns and moves of the middle layers can always be expressed using just the letters F, R, and T—the letters B, L, and D are not necessary.

(f) We also use 90° turns of the cube as a whole, in any direction. For instance, $F_c = F_s F_m$ means the tilting of the whole cube to the right and $F'_c = F'_s F'_m$ means the tilting of the whole cube to the left. The movements $R_c = R_s R_m$, $R'_c = R'_s R'_m$, $T_c = T_s T_m$, and $T'_c = T'_s T'_m$ are defined similarly.

(g) We also use 180° turns of the whole cube: $F^2_c = F^2_s F^2_m$, $R^2_c = R^2_s R^2_m$, and $T^2_c = T^2_s T^2_m$.

If a subsequence consisting of two or more moves is repeated immediately after itself in a sequence, the repeated subsequence will be written only once, in parentheses, and the number of repetitions will be used as the exponent of this parenthesis. Thus the sequence FRTR'T'RTR'T'RTR'T'F' can also be written as $F(RTR'T')^3F'$.

For the sake of clarity, important segments of longer sequences may be separated by the sign · or ×. The use of these symbols is motivated, not so much by the length of the sequence but by the fact that the separated segments of these sequences are usually worth independent study and occur as independent sequences else-where. If necessary, we shall also use the product signs to serve as a hyphen, when we have no room for the whole sequence in the line where it begins.

Where the Magic Domino is concerned, we omit the superscrip

2 after the capitals F, B, R, and L, since only 180° turns of these faces are allowed in this case. We shall underline such letters to emphasize that only 180° turns of these faces are permitted.

The length of a sequence in our tables can be calculated in three different ways.

(a) Half turns and quarter turns are considered to be equivalent. To calculate the length we count turns of faces. A turn of a middle layer is counted as two turns.

(b) Again, half turns and quarter turns are considered to be equivalent, but we count turns of faces or of middle layers, in other words, the turn of a middle layer is equivalent to the turn of a face.

(c) Half turns count double, and we count turns of faces.

In the tables we give the length defined in (a); lengths defined in (b) and (c) will be given in parentheses, but only when they differ from the length defined in (a). One can always decide whether a value in parentheses refers to (b) or (c): if it refers to (b), it is smaller than the length given before the parentheses; if it refers to (c), it is bigger. For example, $F_m^2 = F_s^2 F_c^2$ has length 2(1, 4). [Note that whole cube moves are not counted. This is because each turn in (a) is thought of as a combination of moving the whole cube to a convenient orientation and then carrying out the written move — DS.]

3.2 Restoring the magic cube

3.2.1 *A survey of restoring algorithms*

Masters and enthusiasts of the Magic Cube have worked out many different algorithms to restore the cube in a short time. By a 'short time' we mean two or three minutes, but this supposes a great deal of practice with the algorithm. There are a large number of current methods, and all of them starting by solving a suitable simple sub-problem first. The best known algorithms can be classified into four main groups depending on the subproblem that is solved first. One can restore the corners first, ignoring the edges until all corners have been moved into place. Restoring the edges begins when all the corners are in place, and one must be careful not to confuse the corners again. Of course, the reverse strategy, starting with the edges first and then continuing with the corners, can also be applied. This gives two different types of restoring method. A third method begins with unscrambling a $2 \times 2 \times 3$ block. The fourth main method starts by making one layer correct. The colours on the edges of this layer must correspond with the colours of the centres of the other faces. In other words, we are restoring not just the face but the whole layer, that is, a $1 \times 3 \times 3$ block, making each of its five visible sides of one colour only.

These four types can be classified into subtypes depending on the next stage. If we restore a single layer first, then the next stage can restore the remaining corners, or the remaining edges, or the middle layer parallel to the restored layer. The last of these can be thought of as restoring the cube by floors, starting at the ground floor and finishing at the top floor.

The main aim of the restoring algorithms mentioned so far is unscramble the cube in the shortest possible time, since cubing competitions are judged on the time (not, for example, on the number of turns) taken by the competitors to unscramble the cube. A different kind of challenge is to restore a scrambled cube or create given patterns in as few steps as possible, disregarding the time taken. We shall return to these questions later.

3.2.2   The description of a restoring algorithm

We hope that anybody, even a beginner, will be able to restore the cube from any scrambled pattern if they follow the instructions given in this section. It is relatively easy to learn the algorithm described here and if you practise it several times, it will enable you to restore the cube in one or two minutes, using only remembered processes.

The process we suggest is the following. First restore the four edges of the top face. Then restore the four corners of the same face. Having done this, turn the cube over, so that the down face is now the top. Restore the four edges of the horizontal middle layer, then the four corners of the present top face, and finally its four edges. This is a restoring by floors method, with the slight modification that, while restoring the first layer, we keep it on top in order to see exactly what we are doing; when it is restored we turn the cube over, and from this point the restoring process goes from the bottom up.

It does not seem necessary to suggest a method for restoring the first four edges, since the task is really very simple. We trust that every reader will be able to find his own way of carrying this out after some thought and practice.

Once the edges of the top face have been restored, the algorithm continues by restoring the corners. The easiest case occurs when a corner cube, whose correct place is on the top, is in the down layer, with the facelet that must be moved to the top (that is, the facelet with the same colour as the top centre) positioned on the side of the down layer. In this case turn the whole cube, until the corner cubicle which is the correct place for the cube in question moves to the front right position of the top layer. Suppose the corner cube in question is not in the front right position of the down layer. Then turn the down layer until it gets into this position. Performing either FDF′ or R′D′R in accordance with Figure 3.1, the corner cube in question will then move to its correct position in the correct orientation, without disturbing any other top piece.

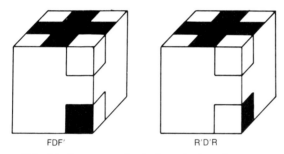

FDF′   R′D′R

Figure 3.1 Moving a corner facelet from the down layer edge to the top

But what do we do, if there is no corner that can be restored in this way? We shall see that this is an awkward situation and it occurs when the colour of the top centre cannot be found on the side of the down layer. In this case take any corner cube which wants to be on top. There are three possibilities for this corner cube, and one of these can occur in two different ways: (a) the cube lies in the down layer with the colour of the top centre on its down facelet; (b) it lies on the top layer, with the correct colour on its top facelet, but it is in an incorrect position; (c) and (d) it lies on the top layer but with one of the two incorrect colours on its top facelet. (In the last case it does not matter whether the cube is in its correct position or not.) Now turn the whole cube, until the corner-piece in question moves to the front right position of top or down. Then perform the appropriate sequence of Figure 3.2. Any top cubes which have already been restored will remain unmoved, and the top

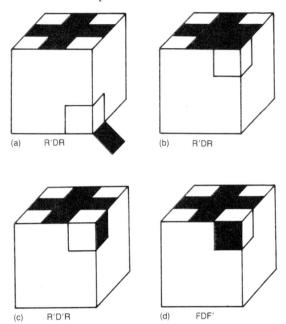

Figure 3.2   Moving an awkward corner facelet to the down layer edge

colour will appear on the side of the down layer. Thus, we can now restore our corner by the method shown in the previous paragraph.

Our present state of knowledge enables us to restore one layer. Having done this, turn the cube over so that the top layer becomes down. Our next move is to restore the horizontal middle layer.

To restore the middle layer, the simplest method is the following. Suppose there is an edge cube on the top layer which does not contain the colour of the top face. If such an edge cube exists, it will be moved to its correct place in seven or eight turns. This edge cube has two facelets, one on the top and one on a lateral face. If the colour on the lateral facelet of this cube does not coincide with the centre of the lateral face, then we turn the top face until they do coincide. Figure 3.3 shows the two possible situations and the appro priate sequence will move this edge cube into its correct position.

But what should be done if one can find no more edges that can be put into their positions using this method? In this case select any edge cube that belongs in the middle layer but is either in the wrong edge cubicle in the middle layer, or is in its correct place, but reversed. Turn the whole cube until the edge in question occupies

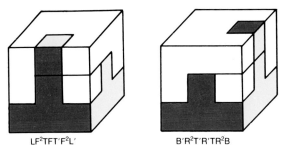

LF²TFT'F²L'   B'R²T'R'TR²B

Figure 3.3 Restoring the edges in a middle layer

the front right position of the middle layer, and apply either of the sequences described in Figure 3.3. (It makes no difference which process you use here.) As a result, the edge cube that stood in the front right position of the middle layer will move to the top layer, and can now be carried to its correct place using the method shown in the previous paragraph. By following these moves the middle layer can be restored.

Next we need to restore the top. Our strategy is as follows: first, we put the corners in their places ignoring their orientation; then we twist them to the correct orientation; next we position the top edges; finally we orientate them.

The processes for restoring the top layer are best illustrated on the so-called distorting mirror or expanded figures. These enable us to see the top *and* all four sides. As our sequences do not cause any change in the two lower layers, our expanded figures will show the upper third of the sides only. We will colour only those squares which are the same colour as the top centre. If the aim of the sequence is just to swap two cubes, the figure will not be coloured at all. If the figure is coloured, then the process given with it will correctly orient its corners, ignoring edges, or, later, will correctly orient its edges, leaving the corners fixed.

To restore the corners of the top we suggest the following process. First, try to get two or more corners of the top in their correct position by turning the top layer. This can always be done, and with luck all four corners will be put in their correct position in this way. If we are not so lucky then we will have to swap two neighbouring corners or two opposite ones. Figure 3.4 shows both cases and sequences to produce the desired swaps. If all the top corners are in place, then either all of them have the correct orientation, or

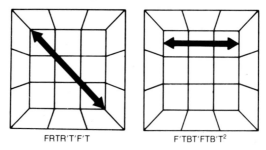

FRTR'T'F'T   F'TBT'FTB'T²

Figure 3.4   Swapping two top corners

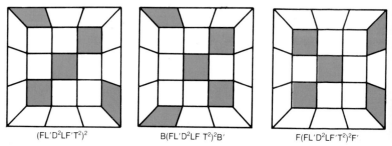

(FL'D²LF'T²)²   B(FL'D²LF T²)²B'   F(FL'D²LF'T²)²F'

Figure 3.5   Twisting two top corners

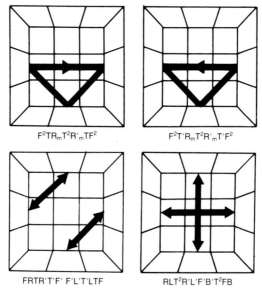

F²TR_mT²R'_mTF²   F²T'R_mT²R'_mT'F²

FRTR'T'F' F'L'T'LTF   RLT²R'L'F'B'T²FB

Figure 3.6   Positioning the top edges

else two, three or four of them are incorrectly oriented. Figure 3.5 shows three sequences that twist two corners.

Each of these sequences twists two corners (the second and the third one are one-step conjugates of the first). If we have three or four improperly oriented corners, then we can twist them to the proper orientations in twos, in two steps. (If we have three badly oriented corners, then, in the first step, we twist one of them into the right orientation, and another one into another, wrong, orientation.)

Finally, the edges of the top can be moved to their correct positions by the sequences given in Figure 3.6 and can be given their correct orientation using Figure 3.7. If two opposite edges, or all four edges are in the wrong orientation, then orientation is finished by using the sequence shown in Figure 3.7 twice (the former case can also be reduced to Figure 3.7 by conjugation).

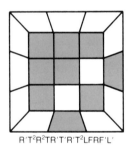

R'T²R²TR'T'R'T²LFRF'L'

Figure 3.7   Flipping two edges

3.2.3   How can you restore the cube more quickly?

The average time for restoring a cube can be improved by learning other sequences, and by modifying and amplifying the basic method introduced in the previous section. Of course, the more tricks we wish to use, the more sequences we have to learn and must be able to apply when needed.

The fast algorithm that we describe next is based on the same outline strategy as the one described in the previous section. It is a restoring-by-floors, with a slight modification—we do not orient the middle layer edges when we put them in place.

To restore the first four edges more quickly, we just point out that it is not necessary to put them into their places; all we need is to put them in the correct cyclic order around the top centre. That is, while restoring the top edges, we may turn the top layer as convenient for getting more top edges on top. But we must put them in the correct position relative to each other. Once all the top edges are in the correct order on the top layer, they can be put in their correct positions in one single turn of that layer.

Restoring the corners of the top layer can be speeded up by using the sequences R′DR, R′D^{2}R, FD′F′ and FD^{2}F′ as well as those pictured in Figure 3.1. To get acquainted with them one should note which corner cube is moved to the top front right position by the individual move sequences, and in what orientation. To understand and remember them, it is best to perform the reverse sequences to the start position. Then you can see exactly what the sequence does.

We modify the method for the middle described in the previous section as follows. If an edge of this layer stands in its correct position, but in the wrong orientation, then we postpone its orientation until the top is finished. We can use the sequences F^{2}T$_s$′R^{2}T$_s$, (F^{2}T^2)^3, F^{2}T$_s^2$B^{2}T$_s^2$ or FBRLT^{2}R′L′F′B′T^2, for rearranging edges within the middle layer. The first, second and third sequences, applied to other layers of the cube, are shown in our tables as sequences 3.8.9, 3.9.38 and 3.9.4; the fourth one is a conjugate of sequence 3.2.4.

The top can be restored in two stages instead of four. To do this, however, we must know all the sequences of Table 3.1 and most of the sequences of Table 3.2. The correct sequence in Table 3.1 moves the corners to their places in their correct orientation. If all the edges of the middle layer are correctly oriented, then the appropriate sequence from Table 3.2 moves the top edges to their places in their correct orientation.

Finally, if there are wrongly oriented edges in the middle layer *and* in the top layer, this is best done by the sequences 3.6.1a and 3.6.2a. Combining these two sequences in various ways we can flip two, four, six or even eight edges into their places in all possible situations. If, for instance, we apply the sequence 3.6.1a, do F$_c^2$ to tilt the cube and then apply the sequence 3.6.1a again—then the four edges of the F face will be flipped. If we simply perform the sequences 3.6.1a and 3.6.2a one after the other, then the front left edge and the front right one will be flipped. It is worth knowing that flips of any four edges can also be reduced to these two sequences by conjugation. With this, our modified algorithm is complete.

To end this section, here are a few final remarks: when using the sequences in Table 3.1, one should note that, when two opposite corners have to be changed over, it does not matter which pair of opposite corners is actually swapped. If, by chance, the correct pair is swapped, then a half turn of the top face will bring all corners back to their correct position at once. This reduces the number of cases involving exchanges of opposite corners down to eight. We first rotate the whole cube around its vertical axis and then perform the sequence of moves given in the appropriate illustration. A further remark is that, if a sequence in Table 3.1 is finished by a turn of the top face, it is usually easier and also more practical to memorize it without this last turn.

It may seem peculiar that we give the same pattern four times, in Table 3.2 numbers 3.2.28–3.2.31. What we want to show is that whenever this pattern occurs on top, we can start restoring it using the same sequence. The same can be said of the pattern belonging to 3.2.32, but here we do not repeat the illustration.

3.2.4. *Strategies using a small number of moves*

In the following we shall sketch two algorithms that guarantee a solution in a small number of moves. Both of them use an extremely large number of alternative sequences (several hundred), and are very complicated, so they are not a quick method for restoring the cube. One of these methods is a 'throw and catch' method that has been worked out by a Cambridge group. The expression refers to the second phase of the method, which consists of three parts.

(a) We pick up a face and restore its four edges. This part of the strategy requires 13 turns at the most, as can easily be proved. However, I would like to mention that Bill Jackson and Katalin Fried have informed me that this subproblem can always be solved in ten, or even eight turns. I must admit, that I have not yet verified this result.

(b) The four corners of this face and the middle layer are restored simultaneously. The maximum number of turns required in this phase is $4 \times 13 = 52$. This has been improved to $4 \times 11 = 44$ turns only, but the improved form is too lengthy to be described here.

(c) The remaining layer is restored using the fast method mentioned in the previous section. This restores the top in at most

31 turns. By using a larger table this number can be reduced to 27, but even this number seems unsatisfactory.

It is part (b) that makes the Cambridge method different from every other. This phase, divided into four parts, is based on a really ingenious idea. They restore a corner and its neighbouring edge in the middle layer simultaneously. In most cases, this is done in two steps, the first step being a 'throw', the second a 'catch'.

The detailed description, and the figures included below, relate to a modified version of the original Cambridge method. It has been worked out by using the Cambridge group's idea.

If the corner and the edge to be simultaneously restored are both in their correct places but either or both of them is incorrectly oriented then, exceptionally, no throw or catch operations are carried out. There are five exceptional cases, and the sequences of Figure 3.8 can be applied to them. These sequences will rotate the front right down corner or the edge above it or both. There will be no other change in the down layer or the middle layer. The cubes of the top layer will be mixed, but for the time being we are not worrying about this layer. Starting with any of the patterns in Figure 3.8

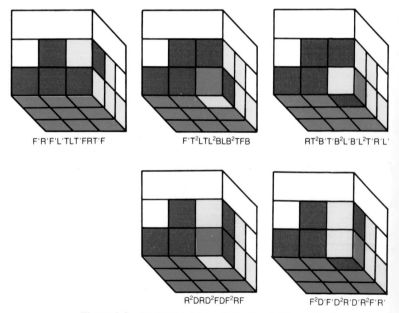

F'R'F'L'TLT'FRT'F   F'T²LTL²BLB²TFB   RT²B'T'B²L'B'L²T'R'L'

R²DRD²FDF²RF   F²D'F'D²R'D'R²F'R'

Figure 3.8   Exceptional cases in the Cambridge method

and performing the given sequence, the down right front corner and the front right edge will be correctly oriented.

If, on the other hand, one or both of the corner and edge in question are not in their correct position, then we need to perform first a throw operation and then a catch operation. Sometimes the throw operation can be omitted and we can start straight away with the catch operation. This is the case when the corner of the pair to be restored is in the down layer and the edge is in the top layer. (This situation will occur, on average, in one case in every four.) In all other cases, when the corner of the pair in question is on the top layer or its edge is on the middle layer and at least one of them is in an incorrect position, we *must* begin with a throw operation. This operation moves the two elements of the pair in different directions—they are 'flying apart'. The corner to be restored will land in the down layer and the edge in the top layer. They are now suitably placed for 'catching', that is, forcing them into their correct positions in the right orientation.

The instructions of Figure 3.9 show how the throw operation can be performed. After the corner is in the down layer and the edge is in the up layer, but their exact location is of no importance, so the illustrations do *not* indicate this. Instead they show their initial positions (before the throw and catch operations) and the positions they want to be in (after the two operations). The throw operation consists of performing the sequence given at the right of the illustration representing the overall problem. Performing this sequence does *not* achieve the change of positions shown on the illustration, but prepares the pieces for a catch operation.

In Figure 3.10 which shows the catch operations, we only have to consider these cases where the corner to be restored is to be carried to the front right position of the down layer, and the edge to be restored is to be carried to the front right position from the top front one. Any other case can be reduced to this situation by rotating the whole cube around its vertical axis, and/or turning the top layer.

Each diagram for the catch operation has a group of six sequences. The sequences of each group produce the same change of position, shown in the diagram, but produce different orientations of the corner and edge in question. As an exercise, we shall leave it to the reader to find the appropriate sequence of the group for various situations. To solve this exercise, it is easiest to first perform the inverse of the catch sequences.

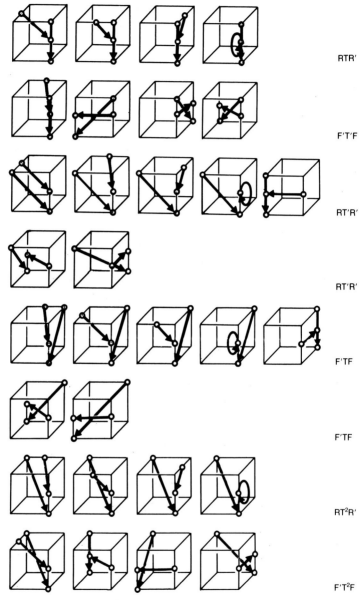

RTR'

F'T'F

RT'R'

RT'R'

F'TF

F'TF

RT²R'

F'T²F

Figure 3.9   The throw operation

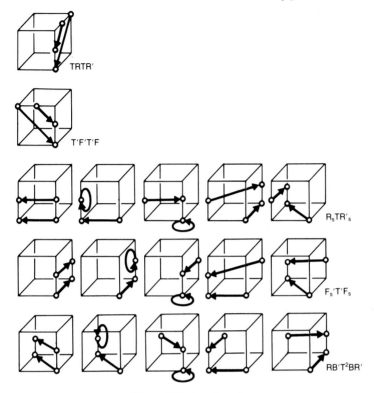

TRTR′

T′F′T′F

R_sTR′_s

F_s′T′F_s

RB′T²BR′

Figure 3.9 (*continued*)

Note that the sequences of Figure 3.10 do not include the initial turn of the top layer.

At present the algorithm using the smallest number of moves is Thistlethwaite's. This method differs from all others in every way. The algorithm is based on the mathematical structures called groups. It starts with the group of all possible patterns that can be reached from the start without taking the cube apart, then it is limited to smaller and smaller subgroups, the last of them being the one-element subgroup. (The one-element group is the one containing the start only.) The groups are defined by progressive restrictions. In the stages of the algorithm quarter turns of the front and back faces, then of the right and left faces, and finally of all faces are prohibited, and only half turns of these faces are allowed. The first stage of Thistlethwaite's algorithm consists of seven moves at

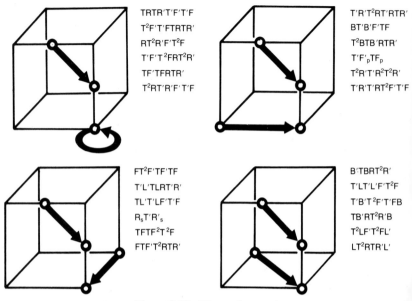

TRTR'T'F'T'F
T²F'T'FTRTR'
RT²R'F'T²F
T'F'T²FRT²R'
TF'TFRTR'
T²RT'R'F'T'F

T'R'T²RT'RTR'
BT'B'F'TF
T²BTB'RTR'
T'F'$_p$TF$_p$
T²R'T'R²T²R'
T'R'T'RT²F'T'F

FT²F'TF'TF
T'L'TLRT'R'
TL'T'LF'T'F
R$_s$T'R'$_s$
TFTF²T²F
FTF'T'RTR'

B'TBRT²R'
T'LT'L'F'T²F
T'B'T²F'T'FB
TB'RT²R'B
T²LF'T²FL'
LT²RTR'L'

Figure 3.10   The catch operation

the most. Whatever the starting pattern, we can, in not more than seven moves, get a pattern such that it can be restored without use of quarter turns of front or back. In the second stage, we get, in at most 13 moves, a pattern that can be restored without using quarter turns of F, B, R, or L. In a further 15 moves, we can achieve a pattern that can be restored by half turns only, and this can be restored in another 17 turns. Adding up these numbers we see that the algorithm restores the cube from any scrambled position in 52 moves at the most. [Work by students of D. E. Knuth has shown that the last stage can be done in 15 moves, reducing the total to 50 moves at most—DS.]

3.3.   Restoring the Magic Domino

Restoring the Magic Domino is both easier, and more difficult than restoring the Magic Cube. It is easier because the number of elements is smaller. On the other hand, it is more difficult, since quarter turns of the sides are not permitted. But, on balance, it is really an easier problem. For instance, the Domino can be restored using only the last two stages of Thistlethwaite's algorithm, since all the patterns of the Domino belong to the smaller group of patterns that can be restored by half turns of the sides and quarter and half turns of the top and down faces. This means a maximum number of $15+17=32$ moves. [Or $15+15=30$, using the result of the last note—DS.]

Now we present another restoring algorithm for the Domino. This one has three stages. First we collect the white corners around the white centre and the black ones around the black centre; then we restore the desired order of both white and black corners; finally we restore the edges by double swaps. Note that there are several possible start patterns that one might want to restore the Domino to.

We shall not go into detail as to how to collect the white (or black) corners around the white (or black) centres since this can be done fairly easily. If, after having done this, the order of the corners relative to each other corresponds to the desired pattern on both faces, we can continue by restoring the edges. If it does not, then two corners must be swapped on one or both faces. By a suitable turn of the square-shaped faces we can produce one of the cases illustrated in Figure 3.11. The figure also gives the relevant move

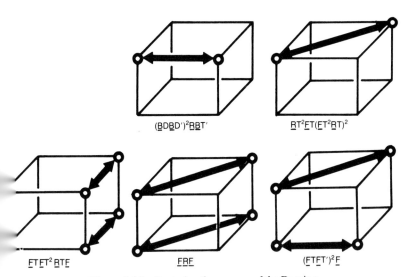

Figure 3.11   Restoring the corners of the Domino

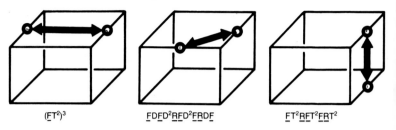

$$(\underline{F}T^2)^3 \qquad\qquad \underline{F}D\underline{F}D^2\underline{R}\underline{F}D^2\underline{F}\underline{R}D\underline{F} \qquad\qquad \underline{F}T^2\underline{R}\underline{F}T^2\underline{F}\underline{R}T^2$$

Figure 3.12   Restoring the edges of the Domino

sequence. Only the edges remain to be restored. This can be done by applying the three move sequences shown in Figure 3.12.

3.4   Instructions for use of the tables

Our tables present a large selection of interesting sequences for the Magic Cube, the Magic Domino and the small ($2 \times 2 \times 2$) cube.

The sequences in our tables are usually accompanied by illustrations and all of them have a serial number preceding the text of the sequence. The numbers to the right of the sequence give the number of moves in three different ways, as described at the end of section 3.1.

Sometimes we give two or three alternative solutions to the same problem. This usually happens when one solution takes less time to perform, but the other takes fewer moves, or if, for various reasons, we cannot decide between the solutions. Sequences achieving the same result with different moves are denoted by the same number and differentiated by small letters. Note that the effects on the cube (Tables 3.2 to 3.9), the Domino (Table 3.10) and the small ($2 \times 2 \times 2$) cube (Table 3.1), brought about by these different sequences must be identical.

Tables 3.1 to 3.4 give one sequence for each important top transformation. There is an essential difference between Table 3.1 and the other three, namely that in Table 3.1 we do not consider the position and orientation of the edges—they can move arbitrarily. Thus Table 3.1 can also be interpreted as a table relating to the small cube, which has no edges. When we discussed transformations earlier we used expanded figures to illustrate them. Here, in Tables 3.1 to 3.4, we use a simplified version of these figures. They take up less room, and still contain all the necessary information. They can even show the movement of the top edges. The form of these simplified, schematic illustrations is basically a square or a triangle. The four (three) vertices of the square (triangle) represent the four top corners of the cube (or three of them). The top edges are imagined to be at the mid-points of the four sides of the square. (They are not shown on triangular diagrams.) The facelet of each corner and edge cube that has the top colour, is marked by a small line sticking out of the square (triangle). If the small line is missing, then the facelet is in its correct position. The exchange of two opposite top corners is represented by connecting the two corresponding points. Since two neighbouring top corners are connected anyway, their exchange is represented by a second line parallel to the corresponding side of the square. The cyclical shift of three cubes is represented by connecting the three corresponding points; the direction of the cycle is represented by an arrow. The sequences given in Tables 3.1 to 3.4 give the changes shown in the corresponding illustrations. Most of these sequences are considered for restoration of the Magic Cube; the small rods therefore represent, not the pattern produced by the sequences applied to start, but the pattern which is restored to start by the sequence. In the case of Table 3.1, this means that the top centre and top corners will become the same colour. In the case of Tables 3.2 to 3.4, the entire face will become a solid colour. If one wants to know how the pattern shown by the illustrations can be reached from the start then the inverse sequence must be performed. One should note that the inverse sequence inverts the direction of cycles of three. In some cases, for instance in all the sequences of Table 3.4, the results of the sequence and of its inverse are identical, that is the sequence is its own inverse. It should also be mentioned that Table 3.4 does not contain all possible cases of this, but only those that can be produced in less than 20 moves. If the reader feels so inclined he may try to extend our list.

To complete Table 3.5 one has to add the top sequences 3.2.1, 3.2.2, 3.3.1, and 3.3.2b.

Sequences 3.6.1a and 3.6.2a have already been discussed. The last four sequences of Table 3.6 flip four, eight or 12 edges simultaneously. In other words, these sequences flip all edges of one, two or three layers. The sequence 3.2.3, discussed earlier, is also related to these sequences. In the illustrations to Table 3.6 the position of the flipped edges are indicated by spots.

To complete Tables 3.7 and 3.8 one has to add the top sequences, 3.3.30 to 3.3.47 and 3.2.14 to 3.2.21 respectively. We have not indicated the orientation of the corners, only their change of position. Thus an illustration can have different sequences. The sequences given in Tables 3.7 and 3.8 all produce the indicated movement of pieces, but with different orientations. [For corner tri-cycles, there are nine possible orientations, for edge tri-cycles, there are four. If the tri-cycle is symmetrically placed, then only five and two of these possibilities need to be given. See 3.7.1 to 3.7.5 and 3.8.1 to 3.8.2 —DS.]

Many regular patterns on the cube can be described as the movement of a group of cubes relative to the rest of the cubes. This motion can be a 180° or a 90° rotation about a face axis, that is, around a straight line connecting the centres of two opposite faces (compare sequences 3.9.1 to 3.9.14 and 3.9.15 to 3.9.18 respectively); a 120° rotation about a corner axis, that is a straight line connecting the two furthest corners (compare sequences 3.9.19 to 3.9.32); a 180° rotation about an edge axis, a straight line connecting the centre of the two furthest edges (compare sequences 3.9.33 to 3.9.36); a central symmetry which changes the positions of opposite edges, either in one layer or in the whole cube (compare sequences 3.9.3 and 3.9.37, respectively) and finally, a reflection in a plane (compare sequences 3.9.4, 3.9.38 and 3.9.39). Sequences 3.9.1 and 3.9.2 result in a 180° rotation of the four top corners and edges respectively; sequences 3.9.3 and 3.9.4 result in a 180° rotation of the four edges in a middle layer; sequence 3.9.5 gives a 180° rotation of four centres relative to the other elements; and so on.

Some of the patterns produced by the sequences of Table 3.9 are very attractive indeed. For instance, the sequence 3.9.5 produces frame (or spot, dot or O), the sequences 3.9.6, 3.9.14 and 3.9.1 produce different cross-shaped patterns, sequence 3.9.10 produce diagonal lines, each on four sides of the cube, and sequence 3.9.2 produces a frame on six sides of the cube. The patterns produce by the sequences 3.9.28 and 3.9.33 have a cross or plus (+) on a faces. The patterns given by the sequences 3.9.29 and 3.9.37 hav an × on all faces of the cube. The pattern of the sequence 3.9.3 has the outline of two 2×2×2 cubes positioned on opposi corners of the Magic Cube, while 3.9.35 has two rings around the opposite corners.

The reader must have noticed, that 90° rotations of a group of elements are less frequent than 180° rotations. This is no accident, since it can be proved that the 90° rotation of less than eight elements is impossible. On the other hand the simultaneous 90° rotation of exactly eight elements, and also that of 16 elements, is possible, but the latter corresponds to a 90° rotation of eight other elements in the opposite direction. (The principle of relativity!)

It is no accident, either, that our examples of central symmetries and reflections in a plane do not change the positions of the corners. We leave it to the reader to prove that this is inevitable.

Table 3.10 lists some sequences for the Magic Domino. Sequences 3.10.1 to 3.10.9 move two or four corners of the Domino, but moving the edges, which is ignored. Sequences 3.10.10 to 3.10.15 move only corners; sequences 3.10.16 to 3.10.27 move only edges.

Table 3.11 lists some sequences that, if we begin at start, will bring the cube back to it like a boomerang. Suppose we write or draw something onto the centres so we can see their orientation. The first four sequences leave the centres in their correct orientation, but the sequences 3.11.5 to 3.11.9 have different effects on the centre orientation. If we have solidly coloured centres, we do not see these change.

Now let us see how the individual sequences of Table 3.11 affect the centres. Sequence 3.11.5 turns both the top centre and the down centre through 180°; sequence 3.11.6 turns only the top centre through 180°. Sequences 3.11.7 and 3.11.8 yield 90° turns, in opposite senses, of two neighbouring and of two opposite centres, respectively. Sequence 3.11.9 turns four centres through 90° simultaneously. Sequence 3.11.5 (3.11.8) operates as follows: before the final turn, the top (down) face is not used; the final T^2 (D') turn returns us to start. This means that the moves before the final one have produced the effect of $T^2(D)$ without ever using that face. That is, a face can be rotated without turning it!

Regarding the orientation of the centre cubes, it is interesting to examine those sequences which move only a small number of elements and see whether they change the orientation of the centres or not. For example, the first three sequences of the group 3.8.9 to 3.8.12 do change them, but the fourth one does not. If the sequence 3.10.26, devised for the Domino, is modified and performed on the cube as $(F^2R^2T^2R^2)^2$, then we can see that the result of this sequence is essentially identical to that of the sequence 3.8.9, but that this time the orientation of the centres is unchanged.

Table 3.1 Top corner sequences (which may move edges)

	Ref.	Sequence	Moves
	3.1.1	$(FL'D^2LF'T^2)^2$	12 (16)
	3.1.2	$BTB'TBT^2B'·B'T'BT'B'T^2B$	13 (16)
	3.1.3a	$BTB'TBT^2B'·F'T'FT'F'T^2F$	14 (16)
	3.1.3b	$LT^2L'T'LT'L'L^2T^2LTL'TL$	13 (16)
	3.1.4	$BTB'TBT^2B'T^2$	8 (10)
	3.1.5	$F'T'FT'F'T^2FT^2$	8 (10)
	3.1.6	$BTB'TBT'B'TBT^2B'$	11 (12)
	3.1.7	$F(RTR'T')^2F'$	10
	3.1.8	$LF'L'T'LT^2BT'FTB'T^2L'$	13 (15)
	3.1.9	$FRTR'F·LFL^2TLT'·'$	11 (13)
	3.1.10	$FRTR'T'F'T$	7
	3.1.11	$B'LTFT^2F'TL'B'TB^2$	11 (13)
	3.1.12	$BTLT'L'TB'TBT^2B'T'$	12 (13)
	3.1.13	$F'T'L'TLT'FT'F'T^2FT'$	12 (13)
	3.1.14	$F(RTR'T')^3F'T$	15
	3.1.15	$FRTR'TF'T'FT'F'T'$	11
	3.1.16	$RB'RF^2R'BRF^2R^2T$	10 (13)
	3.1.17	$BTB'TBTF'TB'T'F$	11
	3.1.18	$F'TBT'FT'B'T'BT'B'$	11
	3.1.19	$LFT'RTR^2F'L'FRF'T$	12 (13)
	3.1.20	$LF'L'BLF^2T'F'TL'B'T'$	12 (13)
	3.1.21a	$F'TBT'FTB'·B'T'BT'B'T^2B$	13 (15)
	3.1.21b	$RTBT'B^2R'(BT^2)'B'T'$	12 (15)

Table 3.1 (*continued*)

	3.1.22a	FT'B'TF'T'B·BTB'TBT²B'	13 (15)
	3.1.22b	L'T'B'TB²L(B'T²)²BT	12 (15)
	3.1.23	R'FRTR'T'F'TR	9
	3.1.24	LF'L'T'LTFT'L'	9
	3.1.25	R'T'FTRT'R'F'R	9
	3.1.26	LTF'T'L'TLFL'	9
	3.1.27	RT²RDR'T²RD'R²T	10 (13)
	3.1.28	L'T²L'D'LT²L'DL²T'	10 (13)
	3.1.29	F²DF'T²FD'F'T²F'T'	10 (13)
	3.1.30	F²D'FT²F'DFT²FT	10 (13)
	3.1.31	BT'F'TB'T²L'TLFT²	11 (13)
	3.1.32	R'TL'T'RLT²L'T'LT²	11 (13)
	3.1.33	FR'D'RT²R'DRT²F'T'	11 (13)
	3.1.34	RT²B'D'BT²B'DBR'T	11 (13)
	3.1.35	F'TBT'FTB'T²	8 (9)
	3.1.36	FT'B'TF'T'BT²	8 (9)
	3.1.37	RB'T²BR'B'RT²R'BT'	11 (13)
	3.1.38	L'BT²B'LBL'T²LB'T	11 (13)
	3.1.39	RB'R'BT²BT²B'T	9 (11)
	3.1.40	L'BLB'T²B'T²BT'	9 (11)
	3.1.41	RT²R'T²R'FRF'T'	9 (11)
	3.1.42	L'T²LT²LF'L'FT	9 (11)

Table 3.2 Top edge moves (leaving the corners fixed)

	3.2.1	$F_sR'TF'RT'F'_sLT'FL'T$	14 (12)
	3.2.2	$R'T^2R^2TR'T'R'T'R'T^2LFRF'L$	13 (16)
	3.2.3	$R_s^2B^2R'_sT^2B'T^2R_sB^2R_s^2T$	14 (10, 22)
	3.2.4	$RLT^2R'L'F'B'T^2FB$	10 (12)
	3.2.5	$R'TB'T(R^2T^2)^2R^2TBT'R$	13 (18)
	3.2.6	$BLFT'F'TL'B'LFTF'T'L'$	14
	3.2.7	$FT^2F'T'L'B'T^2BTL$	10 (12)
	3.2.8	$F_s^2TF^2R_s^2B^2R_s^2T'F'_s$	12 (8, 22)
	3.2.9	$FRTR'T'F'\cdot F'L'TLTF$	11 (12)
	3.2.10	$RLF(T^2R^2)^3F'R'L'$	12 (18)
	3.2.11a	$R_mT'TR_mTR_m'T'R_mTR_m'TR_mT$	18 (12)
	3.2.11b	$R_sBR'_sTR_sB'R'_sTR_sBR'_sT$	18 (12)
	3.2.11c	$F_2LFTFT^2F'L'T'B'T^2BLF'L'F^2$	16 (20)
	3.2.12a	$R_m'T'R_mT'R_m'TR_mT'R_m'T'R_mT'$	18 (12)
	3.2.12b	$R_sB'R'_sT'R'_sBR'_sT'R_sB'R'_sT'$	18 (12)
	3.2.12c	$F^2R'F'T'F'T^2FRTBT^2B'R'FRF^2$	16 (20)
	3.2.13	$L'B'R'TRBT^2\cdot LFRT'R'F'T^2$	14 (16)
	3.2.14a	$F^2TR_mT^2R_m'TF^2$	9 (7, 12)
	3.2.14b	$F^2TR'_mF^2R_sTF^2$	9 (7, 12)
	3.2.15a	$F^2T'R_mT^2R_m'T'F^2$	9 (7, 12)
	3.2.15b	$F^2T'R'_sF^2R_sT'F^2$	9 (7, 12)
	3.2.16a	$R_m'TR_mT^2R_m'TR_m$	11 (7, 12)
	3.2.16b	$R_sBR'_sT^2R_sBR'_s$	11 (7, 12)
	3.2.16c	$F'L'F^2D^2B^2R'B^2D^2F'$	9 (14)

Table 3.2 (*continued*)

3.2.17a	$R'_m T' R_m T^2 R'_m T' R_m$	11 (7, 12)
3.2.17b	$R_s B' R'_s T^2 R_s B' R_s$	11 (7, 12)
3.2.17c	$FRF^2 D^2 B^2 LB^2 D^2 F$	9 (14)
3.2.18	$LFTF'L'B'R'T'RB$	10
3.2.19	$LFT'F'L'B'R'TRB$	10
3.2.20	$R'F'T'FRBLTL'B'$	10
3.2.21	$R'F'TFRBLT'L'B'$	10
3.2.22a	$FRTR'T'F'F'L'T'LTF \cdot F^2 TR_m T^2 R'_m TF^2$	19 (17, 22)
3.2.22b	$BLT'FTF'L'B'R'T'F'TFR$	14
3.2.23a	$FRTR'T'F'F'L'T'LTF \cdot T_c \cdot F^2 TR_m T^2 R'_m TF^2$	20 (18, 24)
3.2.23b	$BLT'FTF'L'B'F'L'TLFRT'R$	16
3.2.24a	$FRTR'T'F'F'L'T'LTF \cdot T_c \cdot F^2 T'R_m T^2 R'_m T'F^2$	20 (18, 24)
3.2.24b	$L'B'TR'T'RBLFTRT'R'F$	14
3.2.25a	$FRTR'T'F'F'L'T'LTF \cdot F^2 T'R_m T^2 R'_m T'F^2$	19 (17, 22)
3.2.25b	$L'B'TR'T'RBLRBT'B'R'F'TF$	16
3.2.26a	$F^2 T'R_m T^2 R'_m T'F^2 \cdot T_c \cdot R'_m TR_m T^2 R'_m TR_m$	20 (14, 24)
3.2.26b	$RTR'F'L'T'LFBLFT'F'TL'B'$	16
3.2.27a	$F^2 TR_m T^2 R'_m TF^2 \cdot T_c \cdot R'_m T'R_m T^2 R'_m T'R_m$	20 (14, 24)
3.2.27b	$L'T'LFRTR'F'B'R'F'TFT'RB$	16
3.2.28a	$RLF(T^2 R^2)^3 F'R'L' \cdot T_c^2 \cdot F^2 TR_m T^2 R'_m TF^2$	21 (19, 30)
3.2.28b	$RBLT'L'TB'R'F'L'T'LTF$	14
3.2.29	$RLF(T^2 R^2)^3 F'R'L' \cdot F^2 T'R_m T^2 R'_m T'F^2$	21 (19, 30)
3.2.30	$RLF(T^2 R^2)^3 F'R'L' \cdot F^2 TR_m T^2 R'_m TF^2$	21 (19, 30)
3.2.31	$RLF(T^2 R^2)^3 F'R'L' \cdot T_c^2 \cdot F^2 T'R_m T^2 R'_m T'F^2$	21 (19, 30)
3.2.32a	$RLF(T^2 R^2)^3 F'R'L' \cdot T_c \cdot F^2 TR_m T^2 R'_m TF^2$	21 (19, 30)
3.2.32b	$L'B'R'TRT'BLFRTR'T'F'$	14

Table 3.3 Top corner moves (leaving the edges fixed)

	3.3.1	$(FL'D^2LF'T^2)^2$	12 (16)
	3.3.2a	$BTB'TBT^2B'\cdot F'T'FT'F'T^2F$	14 (16)
	3.3.2b	$R(RF'D^2FR'T^2)^2R'$	13 (18)
	3.3.3	$(RB'R^2B)^2T^2FRF'R'T^2$	14 (18)
	3.3.4	$T^2RFR'F'T^2(B'R^2BR')^2$	14 (18)
	3.3.5	$L'FD^2LF^2D'FT^2F'DF'L'D^2F'LT^2$	16 (22)
	3.3.6	$L(FT'RTR'TF'T')^2L'$	18
	3.3.7	$F^2R_1^2B^2DF^2R_1^2T$	10 (8, 18)
	3.3.8	$FT^2R'D'RT^2R'DRT'B'TF'T'BT$	16 (18)
	3.3.9	$B'TFT'BTF^2TBT'FTB'T^2$	14 (16)
	3.3.10	$RBL'B'R'B^2R^2B'LBR^2B$	12 (16)
	3.3.11	$B^2R^2B'L'BR^2B^2RBLB'R'$	12 (16)
	3.3.12	$L'T'LT'L'T^2LFT^2F'T'FT'F'T^2$	15 (18)
	3.3.13	$B'T'BT'BTB^2TB^2T^2B'T^2$	12 (16)
	3.3.14	$BTB'TB'T'B^2T'B^2T^2BT^2$	12 (16)
	3.3.15	$FR'F'LFRF'L^2B'RBLB'R'B$	15 (16)
	3.3.16	$FRTR'T'F^2L'T'LTL'TLFRT'R'T^2$	18 (20)

Table 3.3 (*continued*)

	3.3.17	FR'F'LFRF'L'F'TBT'FTB'T'	16
	3.3.18	F(RTR'T)³F'	14
	3.3.19	RT²R'T²R'FRF'T²B'T²B'RBR'T²	17 (16, 20)
	3.3.20	(L'TRT'LTR')²T²	15 (16)
	3.3.21	FR'F'LFRF'L²BL'F²LB'L'F²L²	16 (20)
	3.3.22	F'LFR'F'L'FR²B'RF²R'BRF²R²	16 (20)
	3.3.23	BLF'L'B'LFL'FT'B'TF'T'BT	16
	3.3.24	B'R'FRBR'F'RF'TBT'FTB'T'	16
	3.3.25	L'B'R'TRBLFRT'R'F'	12
	3.3.26	R'F'LFRF'L'F²R'FL²F'RFL²F²	16 (20)
	3.3.27	BL'BR²B'LBR²FB²T'B'TF'T'BT	17 (20)
	3.3.28	B'RB'L²BR'B'L²F'B²TBT'FTB'T	17 (20)
	3.3.29	L'TRT'LTR'B'T²B'D'BT²B'DB²T'	17 (20)
	3.3.30a	L'BL'F²LB'L'F²L²	9 (12)
	3.3.30b	F²L²F'R'FL²F'RF'	9 (12)
	3.3.31a	FR'FL²F'RFL²F²	9 (12)
	3.3.31b	L²F²LBL'F²LB'L	9 (12)
	3.3.32a	F'TBT'FTB'T'	8
	3.3.32b	T'R'TLT'RTL'	8
	3.3.33	B'RT²R'BRB'T²BR'	10 (12)

Table 3.3 (*continued*)

	3.3.34	RB'T²BR'B'RT²R'B	10 (12)
	3.3.35a	LT'R'TL'T'RT	8
	3.3.35b	TBT'F'TB'T'F	8
	3.3.36a	F'L'B'LFL'BL	8
	3.3.36b	RB'R'FRBR'F	8
	3.3.37a	LFRF'L'FR'F'	8
	3.3.37b	B'RBL'B'R'BL	8
	3.3.38	L'T²L'D'LT²L'DL²	9 (12)
	3.3.39	FT²FDF'T²FD'F²	9 (12)
	3.3.40a	L'B'RBLB'R'B	8
	3.3.40b	FRF'LFR'F'L'	8
	3.3.41a	L'BDB'T²BD'B'T²L	10 (12)
	3.3.41b	R'F²R'B²RF²R'B²R²	9 (14)
	3.3.42a	FR'D'RT²R'DRT²F'	10 (12)
	3.3.42b	BL²BR²B'L²BR²B²	9 (14)
	3.3.43a	FRB'R'F'RBR'	8
	3.3.43b	L'B'LF'L'BLF	8
	3.3.44	F²DF'T²FD'F'T²F'	9 (12)
	3.3.45a	FT²R'D'RT²R'DRF'	10 (12)
	3.3.45b	B²R²B'L²BR²B'L²B'	9 (14)
	3.3.46a	B²R²B'L²BR²B'L²B'	10 (12)
	3.3.46b	R²B²RF²R'B²RF²R	9 (14)
	3.3.47	L²D'LT²L'DLT²L	9 (12)

Table 3.4 Top moves which swop a pair of corners and a pair of edges

	3.4.1	RBT'B'TBT²B'T'B'R'BT'BT²B'	16 (19)
	3.4.2	LFL'TRT'LT²R'TRT²R'F'L'	15 (17)
	3.4.3	BTB'R'TF'T'F²RF'T'F'TFT²	15 (17)
	3.4.4	R'TRBT'B'R'FR'F'R²T'	12 (13)
	3.4.5	FL'T'L²DF'D'L'FT'F'T²F'	13 (15)
	3.4.6	B'R'L'TRT'LT²R'TRT²B	13 (15)
	3.4.7a	F'TBT'FT·TB'TBT²B'	11 (13)
	3.4.7b	LT²FRTR'F²LFL²T	11 (13)
	3.4.8	LT'RTR'L'T²RTR'T²	11 (13)
	3.4.9	LBTLT'L'B²T'B²L'B'L²T'L'TL'	16 (19)
	3.4.10	R²B'RBTF'TFT²RTR'TR²	14 (17)
	3.4.11	RBRDL'DLD²BDB'D'R²	13 (15)
	3.4.12a	FT'B'TF'T'·T'BT'B'T²B	11 (13)
	3.4.12b	R'TF'L'T'LF²R'F'R²T'	11 (13)
	3.4.13	R'TL'T'RLT²L'T'LT²	11 (13)
	3.4.14	R'B'T'R'TRB²TB²RBR²TRT'R	16 (19)
	3.4.15	L²BL'B'T'FT'F'T²L'T'LTL²	14 (17)
	3.4.16	L'B'L'D'RD'R'D²B'D'BDL²	13 (15)
	3.4.17	R'L²F'RF²L'FR'F'RLF²L²	13 (17)
	3.4.18	B'RB'R'B²T'B'TBLT'L'	12 (13)
	3.4.19	F²RTR²FRF'TLFL'T²F²	13 (17)
	3.4.20	LR²FL'F²RF'LFR'L'F²R²	13 (17)

Table 3.4 (*continued*)

	3.4.21	BL'BLB²TBT'B'R'TR	12 (13)
	3.4.22	F²L'T'L²F'L'FT'R'F'RT²F²	13 (17)
	3.4.23	R'TR'T'B'DB'D'B²R'B'RBR	14 (15)
	3.4.24	LTL'TLT²L'BT²B'T'BT'LTL'T'B'T	19 (21)
	3.4.25	B'D'RTR'DR²T'RTR²T'B	13 (15)
	3.4.26	F'L'BL'B'L²T'L'TLFT'	12 (13)
	3.4.27	RTBT'B'R'FTF'TFT²F'T	14 (15)
	3.4.28	FR'F'RTRT'R²T'RTRB'R'BT	16 (17)
	3.4.29	FRTR'F²LFL²TLT'	11 (13)
	3.4.30	TR'F²LFL²TLT'FR	11 (13)
	3.4.31	LTFT'F'L²T'L²F'L'F²T'F'T	14 (17)
	3.4.32	B'T'R'TRB²TB²RBR²TRT'	14 (17)
	3.4.33	FT²LF'R'FL'F²RF'R'F²RT²F'	15 (19)
	3.4.34	F'LFL'T²L²D'LT²L'DL²BLTL'T'B'T'	19 (23)
	3.4.35	F'T'FRBTB'R'FTF²LFL²TLT'	17 (19)
	3.4.36	FRT'R'F'LF'L'F²T'F'T	12 (13)
	3.4.37	R'T²B'RFR'BR²F'RFR²F'T²R	15 (19)
	3.4.38	LF'L'FT²F²DF'T²FD'F²R'F'T'FTRT	19 (23)
	3.4.39	LTL'B'R'T'RBL'T'L²F'L'F²T'F'T	17 (19)
	3.4.40	L'B'TBLF'LFL²TLT'	12 (13)

Table 3.5 Basic twists and flips not included in Tables 3.2 and 3.3

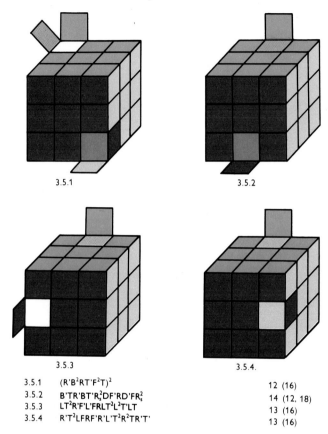

3.5.1

3.5.2

3.5.3

3.5.4.

3.5.1	$(R'B^2RT'F^2T)^2$	12 (16)
3.5.2	$B'TR'BT'R_s^2DF'RD'FR_s^2$	14 (12, 18)
3.5.3	$LT^2R'F'L'FRLT^2L^2T'LT$	13 (16)
3.5.4	$R'T^2LFRF'R'L'T^2R^2TR'T'$	13 (16)

Table 3.6 Flips of 4, 8, 12 edges not included in Table 3.2

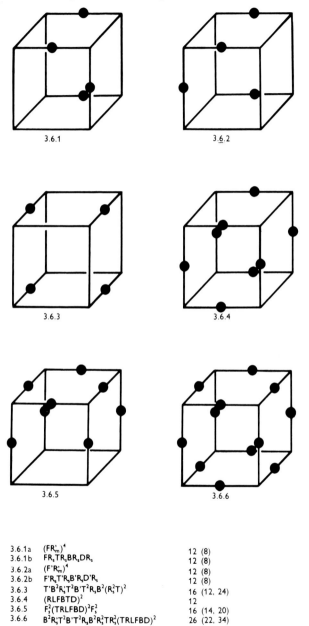

3.6.1a	$(FR'_m)^4$	12 (8)
3.6.1b	$FR_sTR_sBR_sDR_s$	12 (8)
3.6.2a	$(F'R'_m)^4$	12 (8)
3.6.2b	$F'R_sT'R_sB'R_sD'R_s$	12 (8)
3.6.3	$T'B^2R'_sT^2B'T^2R_sB^2(R_s^2T)^2$	16 (12, 24)
3.6.4	$(RLFBTD)^2$	12
3.6.5	$F_s^2(TRLFBD)^2F_s^2$	16 (14, 20)
3.6.6	$B^2R'_sT^2B'T^2R_sB^2R_s^2TR_s^2(TRLFBD)^2$	26 (22, 34)

Table 3.7 3-cycles of corners

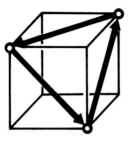

3.7.1	R²BL'B'R²BLB'	8 (10)
3.7.2	L'BLF²L'B'LF²	8 (10)
3.7.3	F'T'RD²R'TRD²R'F	10 (12)
3.7.4	FRD²R'T'RD²R'TF'	10 (12)
3.7.5	FRTR²TL'T'R²TLT²R'F'	13 (16)

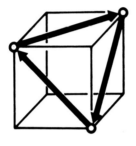

3.7.6	BL'B'R²BLB'R²	8 (10)
3.7.7	F²L'BLF²L'B'L	8 (10)
3.7.8	RTF'D²FT'F'D²FR'	10 (12)
3.7.9	R'F'D²FTF'D²FT'R	10 (12)
3.7.10	R'F'T'F²T'BTF²T'B'T²FR	13 (16)

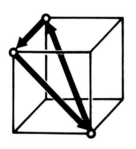

3.7.11	T'LD'L'TLDL'	8
3.7.12	LT'RTL'T'R'T	8
3.7.13	L'D²LTL'D²LT'	8 (10)
3.7.14	TR²T'L'TR²T'L	8 (10)
3.7.15	TBT'F²TB'T'F²	8 (10)
3.7.16	L'B'LF²L'BLF²	8 (10)
3.7.17	B²DBT²B'D'BT²B	9 (12)
3.7.18	B²R'B'L²BRB'L²B'	9 (12)
3.7.19	L'T'L²T'RTL²T'R'T²L	11 (14)

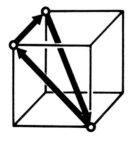

3.7.20	LD'L'T'LDL'T	8
3.7.21	T'RTLT'R'TL'	8
3.7.22	TL'D²LT'L'D²L	8 (10)
3.7.23	L'TR²T'LTR²T'	8 (10)
3.7.24	F²TBT'F²TB'T'	8 (10)
3.7.25	F²L'B'LF²L'BL	8 (10)
3.7.26	B'T²B'DBT²B'D'B²	9 (12)
3.7.27	BL²BR'B'L²BRB²	9 (12)
3.7.28	L'T²RTL²T'R'TL²TL	11 (14)

Table 3.8 3-cycles of edges

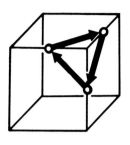

3.8.1	RBLFTF'L'B'R'T'	10
3.8.2	TRR'$_m$T'R$_m$T²R$_m$T'R$_m$R'T'	13 (11, 16)

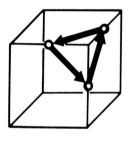

3.8.3	R'D'L'T'F'TLDRF	10
3.8.4	TRR'$_m$TR$_m$T²R'$_m$TR$_m$R'T'	13 (11, 16)

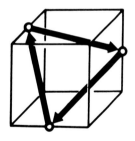

3.8.5	R'TR$_m$T²R'$_m$TR	9 (7, 10)
3.8.6	R'BLFL'F$_s$TF'T'F'R	12 (11)

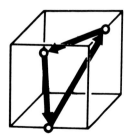

3.8.7	LT'R$_m$T²R'$_m$T'L'	9 (7, 10)
3.8.8	LB'R'F'RF$_s$T'FTFL'	12 (11)

Table 3.8 (*continued*)

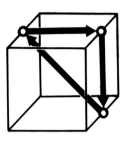

3.8.9	$R^2F_sT^2F_s$	6 (4, 8)
3.8.10	$LFT'F^2L^3B^2D'B'L^2FL'$	11 (16)
3.8.11	$DB'D^2F^2LF'D^2B^2RB'D'$	11 (16)
3.8.12	$RDBTB'TD'RT'R'T'R'$	12

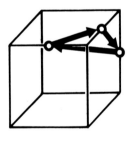

3.8.13	$FT'F'L'B'R'TRBL$	10
3.8.14	$TL'B'R'F'T'FRBL$	10
3.8.15	$L'F^2TR_mT^2R'_mTF^2L$	11 (9, 14)
3.8.16	$R'T^2F'R_mF^2R_mF'T^2R$	11 (9, 14)

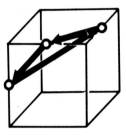

3.8.17	$R'TRBLFT'F'L'B'$	10
3.8.18	$T'BLFRTR'F'L'B'$	10
3.8.19	$BR^2T'F'_mT^2F_mT'R^2B'$	11 (9, 14)
3.8.20	$FT^2RF_mR^2F'_mRT^2F'$	11 (9, 14)

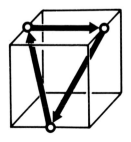

3.8.21	$TR_mT^2R'_mT$	7 (5, 8)
3.8.22	$FL'F^2D^2B^2R'B'B^2D^2F$	9 (14)
3.8.23	$BLFL'F_sTF'T'F'$	10 (9)
3.8.24	$F'T'F'TF_sR'FRB$	10 (9)

Table 3.8 (*continued*)

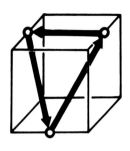

3.8.25	T'R$_m$T^{2}R'$_m$T'	7 (5, 8)
3.8.26	F'RF^{2}D^{2}B^{2}LB^{2}D^{2}F'	9 (14)
3.8.27	B'R'F'RF'$_t$T'FTF	10 (9)
3.8.28	FTFT'F'$_t$LF'L'B'	10 (9)

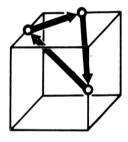

3.8.29	LFR'$_m$F^{2}R$_m$FL'	9 (7, 10)
3.8.30	TR'(T'R')^2(TR)^2	10
3.8.31	(FT)^2(F'T')^{2}F'T	10
3.8.32	LDRTR'T$_s$FT'F'T'L'	12 (11)

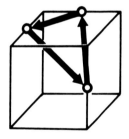

3.8.33	B'R'F$_m$R^{2}F'$_m$R'B	9 (7, 10)
3.8.34	T'F(TF)^2(T'F')^2	10
3.8.35	(R'T')^2(RT)^{2}RT'	10
3.8.36	B'D'F'T'FT'$_t$R'TRTB	12 (11)

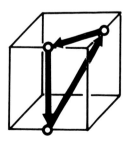

3.8.37	TR^{2}B^{2}L^{2}DL^{2}B^{2}R^2	8 (14)
3.8.38	RF'T$_m$F^{2}T'$_m$F'R'	9 (7, 10)
3.8.39	TF'T'$_t$RT'R'T$_s$F	10 (8)
3.8.40	L'FR'F'R'$_t$DRD'R	10 (9)

Table 3.8 (*continued*)

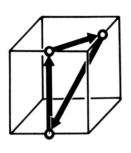

3.8.41	$R^2B^2L^2D'L^2B^2R^2T'$	8 (14)
3.8.42	$RFT_mF^2T_mFR'$	9 (7, 10)
3.8.43	$F'T_s'RTR'T_sFT'$	10 (8)
3.8.44	$R'DR'D'R_sFRF'L$	10 (9)

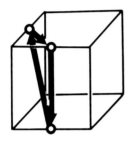

3.8.45	$T'L^2B^2R^2D'R'B^2L^2$	8 (14)
3.8.46	$L'FT_m'F_2T_mFL$	9 (7, 10)
3.8.47	$T'FT_sL'TLT_s'F'$	10 (8)
3.8.48	$RF'LFR_s'D'L'DL'$	10 (9)

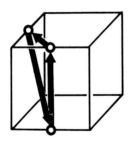

3.8.49	$L^2B^2R^2DR'B^2L^2T$	8 (14)
3.8.50	$L'F'T_m'F^2T_mF'L$	9 (7, 10)
3.8.51	$FT_sL'T'LT_s'F'T$	10 (8)
3.8.52	$LD'LDR_sF'L'FR'$	10 (9)

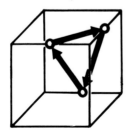

3.8.53	$F^2DF_mD^2F_m'DF^2$	9 (7, 12)
3.8.54	$F'B'L'F_m'L^2F_mL'FB$	11 (9, 12)
3.8.55	$T^2BT_sR'TRT_s'B'T$	11 (9, 12)
3.8.56	$BR'TRBLFT'F'L'B^2$	11 (12)

Table 3.8 (*continued*)

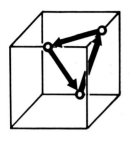

3.8.57	$F^2D'F_mD^2F'_mD'F^2$	9 (7, 12)
3.8.58	$F'B'LF'_mL^2F_mLFB$	11 (9, 12)
3.8.59	$T'BT_tR'T'RT'_tB'T^2$	11 (9, 12)
3.8.60	$B^2LFTF'L'B'R'T'RB'$	11 (12)

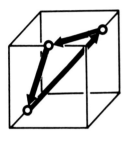

3.8.61	$R^2D'R'_mD^2R_mD'R^2$	9 (7, 12)
3.8.62	$RLBR_mB^2R'_mBR'L'$	11 (9, 12)
3.8.63	$T^2L'T'_sFT'F'T_sLT'$	11 (9, 12)
3.6.64	$L'FT'F'L'B'R'TRBL^2$	11 12

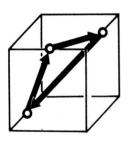

3.8.65	$R^2DR'_mD^2R_mDR^2$	9 (7, 12)
3.8.66	$RLB'R_mB^2R'_mB'R'L'$	11 (9, 12)
3.8.67	$TL'T'_sFTF'T_sLT^2$	11 (9, 12)
3.8.68	$L^2B'R'T'RBLFTF'L$	11 (12)

Table 3.9 Patterns produced by symmetry

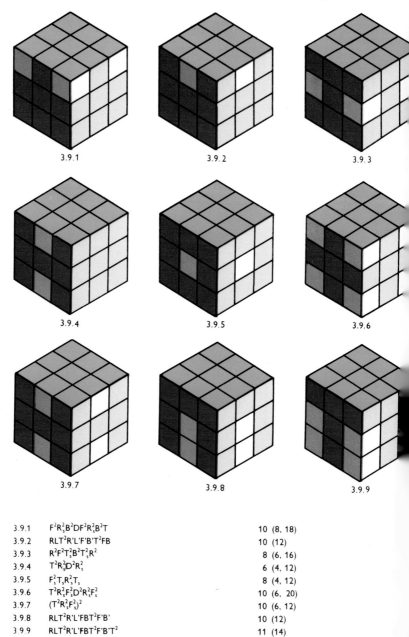

3.9.1	$F^2R_s^2B^2DF^2R_s^2B^2T$	10 (8, 18)
3.9.2	$RLT^2R'L'F'B'T^2FB$	10 (12)
3.9.3	$R^2F^2T_s^2B^2T_s^2R^2$	8 (6, 16)
3.9.4	$T^2R_s^2D^2R_s^2$	6 (4, 12)
3.9.5	$F_s^2T_sR_s^2T_s$	8 (4, 12)
3.9.6	$T^2R_s^2F_s^2D^2R_s^2F_s^2$	10 (6, 20)
3.9.7	$(T^2R_s^2F_s^2)^2$	10 (6, 12)
3.9.8	$RLT^2R'L'FBT^2F'B'$	10 (12)
3.9.9	$RLT^2R'L'FBT^2F'B'T^2$	11 (14)

Table 3.9 (*continued*)

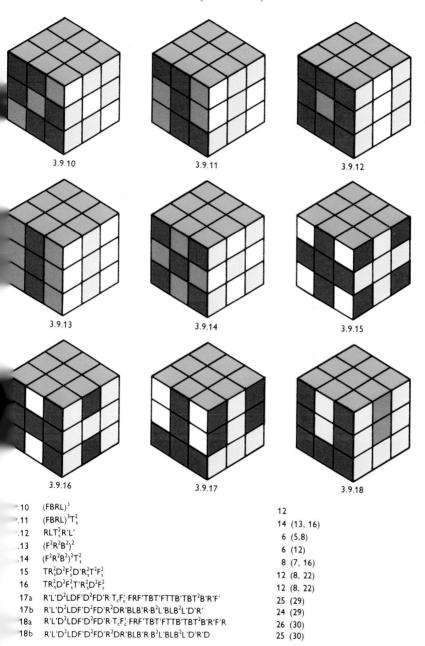

3.9.10   3.9.11   3.9.12

3.9.13   3.9.14   3.9.15

3.9.16   3.9.17   3.9.18

.10	$(FBRL)^3$	12
.11	$(FBRL)^3T_s^2$	14 (13, 16)
.12	$RLT_s^2R'L'$	6 (5,8)
.13	$(F^2R^2B^2)^2$	6 (12)
.14	$(F^2R^2B^2)^2T_s^2$	8 (7, 16)
15	$TR_s^2D^2F_s^2D'R_s^2T^2F_s^2$	12 (8, 22)
16	$TR_s^2D^2F_s^2T'R^2_sD^2F_s^2$	12 (8, 22)
17a	R'L'D²LDF'D²FD'R·T_cF_c·FRF'TBT'FTTB'TBT²B'R'F'	25 (29)
17b	R'L'D²LDF'D²FD'R²DR'BLB'R·B²L'BLB²L'D'R'	24 (29)
18a	R'L'D²LDF'D²FD'R·T_cF_c·FRF'TBT'FTTB'TBT²B'R'F'R	26 (30)
18b	R'L'D²LDF'D²FD'R²DR'BLB'R·B²L'BLB²L'D'R'D	25 (30)

Table 3.9 (*continued*)

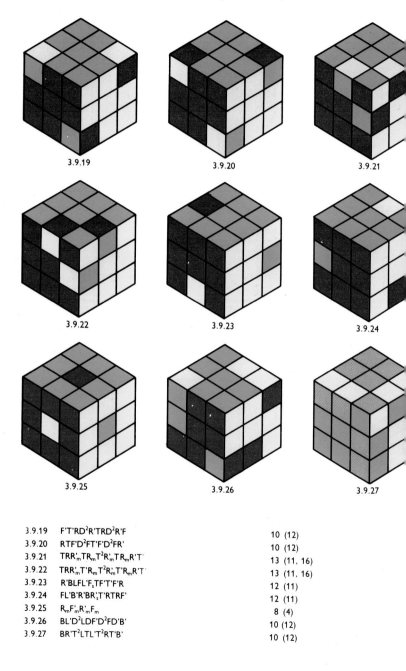

3.9.19

3.9.20

3.9.21

3.9.22

3.9.23

3.9.24

3.9.25

3.9.26

3.9.27

3.9.19	F'T'RD²R'TRD²R'F	10 (12)
3.9.20	RTF'D²FT'F'D²FR'	10 (12)
3.9.21	TRR'ₘTRₘT²R'ₘTRₘR'T'	13 (11, 16)
3.9.22	TRR'ₘT'RₘT²R'ₘT'RₘR'T'	13 (11, 16)
3.9.23	R'BLFL'FₛTF'T'F'R	12 (11)
3.9.24	FL'B'R'BR'ₛT'RTRF'	12 (11)
3.9.25	RₘF'ₘR'ₘFₘ	8 (4)
3.9.26	BL'D²LDF'D²FD'B'	10 (12)
3.9.27	BR'T²LTL'T²RT'B'	10 (12)

Table 3.9 (*continued*)

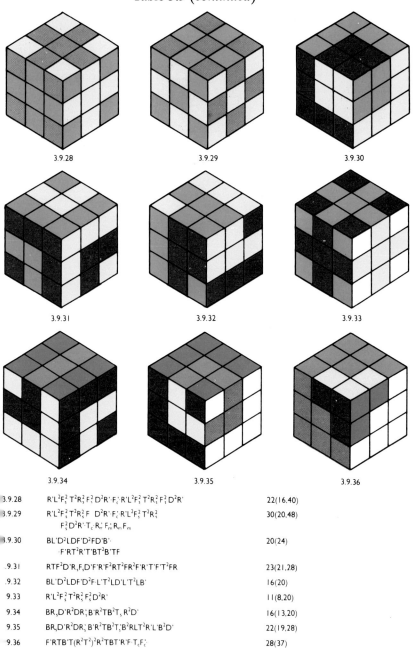

3.9.28

3.9.29

3.9.30

3.9.31

3.9.32

3.9.33

3.9.34

3.9.35

3.9.36

3.9.28	$R'L^2F_s^2\,T^2R_s^2\,F_s^2\,D^2R\cdot F_c\cdot R'L^2F_s^2\,T^2R_s^2\,F_s^2\,D^2R\cdot$	22(16,40)
3.9.29	$R'L^2F_s^2\,T^2R_s^2\,F\ \ D^2R\cdot F_c\cdot R'L^2F_s^2\,T^2R_s^2$ $F_s^2D^2R'\cdot T_c\cdot R_n'\,F_m'R_m'F_m$	30(20,48)
3.9.30	$BL'D^2LDF'D^2FD'B'\cdot$ $\cdot F'RT^2R'T'BT^2B'TF$	20(24)
3.9.31	$RTF^2D'R_sF_sD'F'R'F^2RT^2FR^2F'R'T'F'T^2FR$	23(21,28)
3.9.32	$BL'D^2LDF'D^2F\cdot L'T^2LD'L'T^2LB'$	16(20)
3.9.33	$R'L^2F_s^2\,T^2R_s^2\,F_s^2D^2R'$	11(8,20)
3.9.34	$BR_sD'R^2DR_s'B'R^2TB^2T_sR_s'D'$	16(13,20)
3.9.35	$BR_sD'R^2DR_s'B'R^2TB^2T_s'B^2RLT^2R'L'B^2D'$	22(19,28)
3.9.36	$F'RTB'T(R^2T^2)^2R^2TBT'R'F'T_cF_c'\cdot$ $R^2LT'FRTR'F^2LFL'T R^2$	28(37)

Table 3.9 (*continued*)

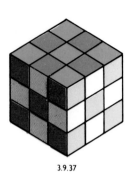

3.9.37   3.9.38   3.9.39

3.9.37	$F_s^2 R_s^2 T_s^2$	6(3,12)
3.9.38	$(F^2 R^2)^3$	6(12)
3.9.39	$F^2 R^2 F^2 R_s^2 F^2 R^2 B^2$	8(7,16)

Table 3.10 Magic Domino sequences

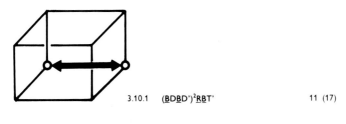

3.10.1 (B̲DB̲D')²R̲BT'   11 (17)

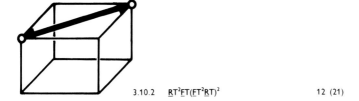

3.10.2 R̲T²F̲T(F̲T²R̲T)²   12 (21)

3.10.3 T²B̲D'(B̲TB̲T')²L̲T²   13 (21)

3.10.4 T²F̲T(F̲T²R̲T)²R̲   12 (21)

3.10.5 F̲D'(F̲TF̲T')²R̲   11 (17)

Table 3.10 (*continued*)

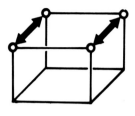

3.10.6 $\underline{R}T^2\underline{F}(T\underline{F}T^2\underline{R})^2$ 11 (20)

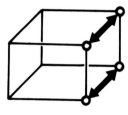

3.10.7 $\underline{F}T\underline{F}T^2\underline{R}T\underline{F}$ 7 (12)

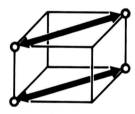

3.10.8 $\underline{F}R\underline{F}$ 3 (6)

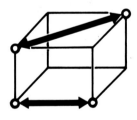

3.10.9 $(\underline{F}T\underline{F}T')^2\underline{F}$ 9 (14)

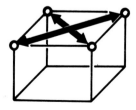

3.10.10a $\underline{F}T^2(\underline{F}RT^2)^2\underline{R}$ 9 (18)
3.10.10b $\underline{F}R\text{,}BDF\underline{R}\text{,}BT$ 10 (8, 18)

Table 3.10 (*continued*)

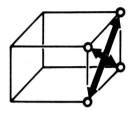

3.10.11 $T^2\underline{RF}(T^2\underline{FR})^2$ 9 (18)

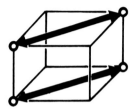

3.10.12 $\underline{F}RT^2\underline{RF}T^2\underline{FR}T^2$ 9 (18)

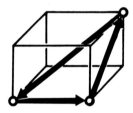

3.10.13 $(\underline{R}T'\underline{L}T)^2$ 8 (12)

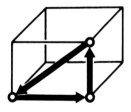

3.10.14 $T'(\underline{R}T'\underline{L}T)^2T$ 9 (14)

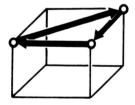

3.10.15 $\underline{F}(RT'\underline{L}T)^2\underline{F}$ 10 (16)

Table 3.10 (*continued*)

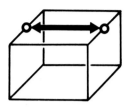

3.10.16 $(\underline{F}T^2)^3$ 6 (12)

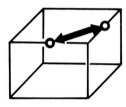

3.10.17 $\underline{F}D\underline{F}D^2\underline{R}\underline{F}D^2\underline{F}RD\underline{F}$ 11 (20)

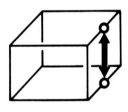

3.10.18 $\underline{F}T^2\underline{R}\underline{F}T^2\underline{F}R\underline{R}T^2$ 8 (16)

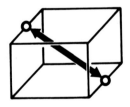

3.10.19 $\underline{R}(\underline{F}T^2)^3\underline{R}$ 8 (16)

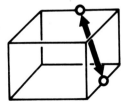

3.10.20 $T\underline{F}T^2\underline{R}\underline{F}T^2\underline{F}R T$ 9 (16)

Table 3.10 (*continued*)

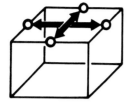

3.10.21 F̲R̲ₛB̲DF̲R̲ₛB̲T'   10 (8, 18)

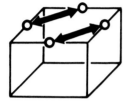

3.10.22 F̲ₛTF̲R̲ₛB̲R̲ₛT'F̲ₛ   12 (8, 22)

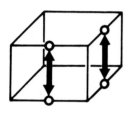

3.10.23 (F̲R̲)³   6 (12)

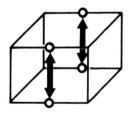

3.10.24 F̲R̲ₛB̲R̲ₛ   6 (4, 12)

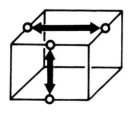

3.10.25 F̲R̲T²F̲R̲F̲T²R̲   8 (16)

Table 3.10 (*continued*)

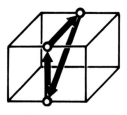

3.10.26 $(\underline{F}R T^2 \underline{R})^2$ 8 (16)

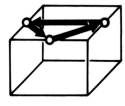

3.10.27 $\underline{F}T(\underline{F}R T^2 \underline{R})^2 T' \underline{F}$ 12 (22)

Table 3.11

3.11.1	$(F_s^2 R_s^2)^2$	8(4,16)
3.11.2	$RTR'F'T^2L'T'LFT^2$	10(12)
3.11.3	$RTR'F'TF'LFL'T'FT'$	12
3.11.4	$(F^2R^2)^6$	12(24)
3.11.5	$F^2R_s^2B^2D^2F^2R_s^2B^2T^2$	10(8,20)
3.11.6	$(RLT^2R'L'T)^2$	12(14)
3.11.7	$R_sF_sT_sR'T'_sF'_sR'_sT$	14(8)
3.11.8	$R_sF_s^2R_sTR_sF_s^2R_sD'$	14(8,18)
3.11.9	$R_sT_sF'_sR_sT_sR'_sF_sT'_s$	16(8)

4

MATHEMATICS

Gerzson Kéri and Tamás Varga

4.1 Mathematical proofs

Mathematicians appreciate Rubik's cube because it presents many challenging problems. The cube, and its simplified versions, the Magic Domino and the $2 \times 2 \times 2$ cube, have mathematical structures which can be analysed by group and combinatorial tools. However, it is possible to prove certain facts about the cube without using complicated mathematics, such as the following.

In the statements (a) to (d), we consider patterns with respect to the centres. That is, we do not distinguish patterns which differ by turning the whole cube. Also, we assume only standard turns—the twisting of a single loose corner is not permitted.

(a) It is *not* possible to exchange a pair of cubes without moving other cubes.

(b) It is *not* possible to twist a single corner without moving a corner.

(c) It is *not* possible to flip a single edge without moving other edges.

(d) The number of different arrangements of colours on the surface of the cube is 43 252 003 274 489 856 000.

(e) The number of colour patterns is 12 times the above number if the cube is taken apart and reassembled. These patterns fall into 12 classes. Turning the faces can take you from a pattern to any pattern in the same class but can never take you out of the class.

(a) *Two cubes cannot be exchanged without moving other cubes*
This proof applies to the corner and edge pieces considered all together. Thus, we do not exclude exchanging two corners and two

146

edges at the same time while the rest of the corners and edges keep their places; Table 3.4 shows this can be done. We are concerned here only with the position of the cubes, not with their orientation.

Take the cube in its start state. Number both the cubes and the cubicles in any order from 1 to 20. Turning the cube changes the order of the cubes but the cubicles are considered fixed. Now read off the numbers of the cubes in cubicles 1, 2, ... 20. If just two cubes were exchanged the list would read, say,

$$1, \mathbf{5}, 3, 4, \mathbf{2}, 6, 7, \ldots \text{etc.}$$

It is clearly impossible if the two numbers, here **2** and **5**, represent a corner and an edge. We shall see that the exchange of two corners or of two edges is just as impossible. The following reasoning is independent of any particular choice of numbers, but will be shown for the exchange of **2** and **5**.

The basic idea is that there is a simple measure of the distance between the initial ascending order and our pattern. This is the number of exchanges of two adjacent entries that is required to get from one to the other. In our example this may be done by, first, moving **2** adjacent to **5**:

$$1, \mathbf{5}, 3, 4, \mathbf{2}, 6, 7, \ldots$$
$$1, \mathbf{5}, 3, \mathbf{2}, 4, 6, 7, \ldots$$
$$1, \mathbf{5}, \mathbf{2}, 3, 4, \ 6, 7, \ldots \quad \text{(two swaps);}$$

then exchanging them:

$$1, \mathbf{2}, \mathbf{5}, 3, 4, 6, 7, \ldots \quad \text{(one swap);}$$

then moving **5** to its correct place:

$$1, \mathbf{2}, 3, \mathbf{5}, 4, 6, 7, \ldots$$
$$1, \mathbf{2}, 3, 4, \mathbf{5}, 6, 7, \ldots \quad \text{(two swaps)}$$

Whenever two elements are exchanged, the number of swaps is always an odd number, $n+1+n$, where n is the number of swaps making them adjacent.

On the other hand, we shall see that a quarter turn of any face gives rise to an *even* number of swaps. A half turn, since even+even=even also produces an even number. Similarly, any sequence of moves results in an even number of swaps, never an odd number, and hence the exchange of two cubes only is not possible.

Consider the quarter turn depicted in Figure 4.1. We are going to

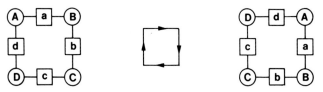

Figure 4.1   When a face is turned eight cubes change place; in how many swaps of two cubes can these changes be brought about?

break this turn up into individual swaps by imagining that we can take out and replace pieces. Whatever the cubes or their numbers, each such swap exchanges one pair of numbers in our list.

Figure 4.2. shows that three swapping steps will move the corners into their new places.

1. Swap A with B. This moves A into its new cubicle.

2. Swap B with C. This moves B into its new cubicle.

3. Swap C with D. This moves both of them into their new cubicles.

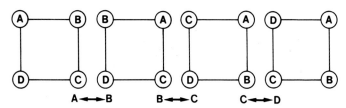

Figure 4.2   The corners can be positioned in three swaps

Figure 4.3 shows that the edges can also be moved into their correct cubicles in three swaps. Three plus three, twice three, is an even number. Each swap of any two elements can be achieved by an *odd* number of swaps of adjacent elements in a list but the sum of *six odd numbers* (i.e. one odd number for each swap) is an *even* number. (The figures show a clockwise quarter turn. The same argument and results hold for an anticlockwise quarter turn.)

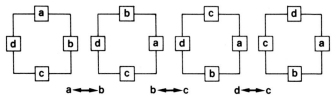

Figure 4.3   The edges can also be positioned in three swaps

However, we have tacitly assumed that two elements exchanged by an odd number of swaps between adjacent elements cannot ever be exchanged by an even number of such swaps. The reasoning which follows avoids this assumption and proves that an even number of adjacent swaps can never be equal to an odd number of them.

Number the cubes and the cubicles from 1 to 20 as earlier. Let c_i denote the cube is the i-th cubicle, for $i = 1, 2, \ldots, 20$. Consider all the quotients

$$\frac{c_k - c_i}{k - i}, \text{ where}$$

$1 \leq k < i \leq 20$. These are:

$$\frac{c_1 - c_2}{1 - 2}, \quad \frac{c_1 - c_3}{1 - 3}, \quad \ldots, \quad \frac{c_1 - c_{20}}{1 - 20},$$

$$\frac{c_2 - c_3}{2 - 3}, \quad \ldots, \quad \frac{c_2 - c_{20}}{2 - 20},$$

$$\ldots$$

$$\frac{c_{19} - c_{20}}{19 - 20}.$$

(There are $19 + 18 + \ldots + 1 = 190$ of them.)

Their product, which we will call Z, will be proved always equal to $+1$.

Exchanging c_k for c_l changes only the values of the quotients containing one or both of the indices k and l. Assuming $k < l$, they will be modified as follows.

(a) For $i = 1, 2, \ldots, k-1$:
$c_i - c_k$ is replaced by $c_i - c_l$ and vice versa.

(b) For $i = l+1, l+2, \ldots, 20$:
$c_k - c_i$ is replaced by $c_l - c_i$ and vice versa.

(c) For $i = k+1, k+2, \ldots, l-1$:
$c_k - c_i$ is replaced by $c_l - c_i$ while $c_i - c_l$ is replaced by $c_i - c_k$.

(d) $c_k - c_l$ is replaced by $c_l - c_k$.

In (a) and (b), only the order of the numerators is changed and this does not change the value of Z.

In (c) an *even number* of numerators are multiplied by -1 and the order of the numerators has changed, but this does not change the value of Z.

In (d) the value of Z is multiplied by -1, since $c_l - c_k$ is -1 times $c_k - c_l$.

Thus any single exchange changes the sign of Z. The rest of our previous reasoning remains; a quarter turn causes an even number of cubes (or numbers) to be exchanged and hence the value of Z remains unchanged. It is therefore impossible to swap two cubes without moving other cubes.

(b) and (c) *No cube can be twisted or flipped on its own.* Now corners and edges are considered separately—twisting one corner and flipping one edge is just as impossible as changing the orientation of only one of them.

We consider the corners first. A corner cannot be twisted, either clockwise or anticlockwise while all other corners remain fixed.

For our purposes the front and the back faces of the cube will be called the main faces. At start, we number the three facelets of each corner cube as follows:

those on the main faces by 0;

those on the other two faces will be numbered so as to have $0, \frac{1}{3}, \frac{2}{3}$ reading clockwise at every corner.

This is shown in Figure 4.4.

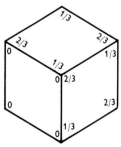

Figure 4.4   Assigning numbers to the facelets of the corners: 0s on the main faces and $0, \frac{1}{3}, \frac{2}{3}$, clockwise at every corner

We shall show that a quarter turn of a face always changes the sum of the numbers on the main faces by an integer, that is, a whole number. Hence, the same is true for a half turn, a complete turn and any sequence of turns. On the other hand, twisting one corner changes the sum of the numbers on the main faces either by $\frac{1}{3}$ or b

$\frac{2}{3}$. Therefore neither a clockwise nor an anticlockwise twist, leaving the other corners fixed, can be brought about by turning the faces.

The fact that a quarter turn of a face changes the sum of the numbers on the main faces by an integer is shown in Figure 4.5. Consider the cube c_1. If its main facelet is 0 before the turn, then it is $\frac{2}{3}$ after the turn. If it was $\frac{1}{3}$, it becomes 0, if it was $\frac{2}{3}$, it becomes $\frac{1}{3}$. So the turn either subtracts $\frac{1}{3}$ or adds $\frac{2}{3}$. The same holds for cube c_3, but at cubes c_2 and c_4, the turn either adds $\frac{1}{3}$ or subtracts $\frac{2}{3}$. So we have an increase by $\frac{2}{3}$ or a decrease by $\frac{1}{3}$ at two corners and an increase by $\frac{1}{3}$ or a decrease by $\frac{2}{3}$ at the other two. The sum of these changes is always an integer.

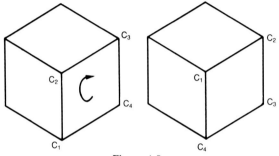

Figure 4.5

Though the figure shows a clockwise turn of a right face, the result is the same if any other face is turned clockwise or anti-clockwise. (Turning a main face does not change the sum of the numbers on the main faces. The change is always 0, which is an integer.)

Now we consider the edge cubes. Let us mark the facelets of each edge piece as shown in Figure 4.6. There is an 0 on one facelet of

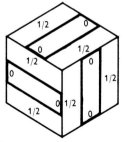

Figure 4.6 Assigning numbers to the facelets of the edges: 0s in the frames, $\frac{1}{2}$ elsewhere; the frames on the unseen faces are situated the same way as ('parallel to') those which are seen

each edge cube and a $\frac{1}{2}$ on the other facelet. Let us also suppose that there are frames round the 0s as indicated by the emphasized lines in Figure 4.6; on opposite faces of the cube these frames will be parallel to each other. If we imagine that these frames remain fixed and that the numbers move when the faces of the cube are turned, then we will prove that a quarter turn of any face and, hence, any sequence of turns, changes the *sum of the numbers in the frames* by a whole number. Flipping a single edge would change this sum by $\frac{1}{2}$, not by a whole number, and it therefore follows that a single edge piece cannot be flipped on its own.

Take, for example, the right face as shown in Figure 4.7. After a clockwise quarter turn the numbers in the frames will have moved out of them, while the numbers which were outside them will now be inside. Look at the edge cube E_1. If it had 0 in the frame on the front face before the turn, it has $\frac{1}{2}$ in the frame on the right face after the turn. If it had $\frac{1}{2}$ before, then it has 0 after. Thus its contribution to the sum of the numbers in frames changes by $+\frac{1}{2}$ or $-\frac{1}{2}$. The same holds for E_2, E_3, E_4. But four such changes is always an integer.

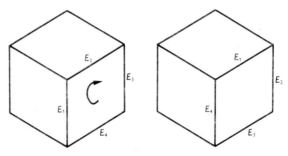

Figure 4.7

(d) *The number of possible patterns on Rubik's Cube, the Domino and the small cube* ($2 \times 2 \times 2$ *cube*). Take the cube apart and then replace seven of its eight corners and ten of its 12 edges. In how many ways is this possible? The first corner can be fitted into one of eight different cubicles, seven places are left for the second, six for the third, . . . two for the seventh. The first two corners (disregarding their orientation) can be replaced in 8×7 different ways, and the number of possibilities is multiplied by 6, . . . making

$8 \times 7 \times 6 \times 5 \times 4 \times 3 \times 2$ possibilities for seven corners. (The product $8 \times 7 \times \ldots \times 2 \times 1$ is denoted 8! and in general, $n \times (n-1) \times \ldots \times 2 \times 1$ is n!) Since the corners have three orientations, each of the above numbers is multiplied by three, which gives $8 \times 7 \times 6 \times 5 \times 4 \times 3 \times 2 \times 3 \times 3 \times 3 \times 3 \times 3 \times 3 \times 3$ or 88 179 840 possibilities ($= 8! \times 3^7$) oriented ways to replace seven corners.

Similarly, the ten edges can be replaced in $12 \times 11 \times 10 \times 9 \times 8 \times 7 \times 6 \times 5 \times 4 \times 3 \times 2 \times 2 \times 2 \times 2 \times 2 \times 2 \times 2 \times 2 \times 2 \times 2$ ($= 12!/2 \times 2^{10} = 12! \times 2^9$) or 245 248 819 200 different ways.

The latter can be freely combined with the $8! \times 3^7$ possibilities for the corners. Therefore the number of possibilities for reassembling seven corners and ten edges is $8! \times 3^7 \times 12! \times 2^9$ or 21 626 001 637 244 928 000. When replacing the remaining three cubes there are two options. The first is to disregard the possibility that the cube might be assembled so that it cannot be restored to start. In this case there are:

three possibilities for the eighth corner; four possibilities for the eleventh edge (two choices for position and two for orientation); two possibilities for the 12th edge (a choice between two orientations).

These again can be freely combined, both with each other and with the possibilities already mentioned. This gives $3 \times 4 \times 2$ times the above number as the number of possible patterns obtained by reassembling the cube:

$$8! \times 3^8 \times 12! \times 2^{12} \text{ or}$$

$$519\ 024\ 039\ 292\ 878\ 272\ 000, \text{ about } 5.2 \times 10^{26}.$$

The second option for the last three cubes is to construct it so it can be restored to start. This requirement determines:

the orientation of the eighth corner in addition to its position — since the sum of the main facelets must be an integer;

the positions of the last two edges — since the number of adjacent swaps required must be even;

the orientation of the last edge — since the sum of the frame face-lets must be an integer.

The only remaining choice is in the orientation of the last edge but one. Therefore the number of possibilities for this second option is twice, rather than 24 times, $8! \times 3^7 \times 12! \times 2^9$, that is $8! \times 3^7 \times 12! \times 2^{10}$ or 43 252 003 274 489 856 000, about 4.3×10^{19}.

This is the number of possible patterns which might be reached from start. By examining any of our algorithms carefully, you can see that you can actually achieve all these patterns.

We can deduce from the above reasoning, that the $8! \times 3^8 \times 12! \times 2^{12}$ different patterns, produced by taking the cube apart and reassembling it, fall into 12 different classes. You can go from one pattern to another in the same class by turning but you can never go between patterns in different classes.

The eight corners and eight edges of the $3 \times 3 \times 2$ domino cannot change orientation. There is no restriction on the number of swaps — any two corners or any two edges can be swapped. So, it would seem that the number of possible states of the domino is 8! (for the corners) times 8! (for the edges) or $(8!)^2$. But the 2×3 faces cannot be distinguished by their centres as can the faces of Rubik's cube. Thus we canot distinguish patterns which differ only by turning the whole domino about its short axis, that is, by applying T_c in Figure 2.80. Thus the number of different patterns is one-quarter of $(8!)^2$.

It is worth examining this argument in several ways. We have already said that we ignore turns of the whole cube, so it might seem that we had already accounted for the T_c turns of the domino. T_c applied to the domino is also produced by T_s, which is a different pattern on the cube. Repeating T_s gives four different cube patterns which are the same domino pattern. Our calculation of $(8!)^2$ was based on having visible centres to distinguish these different patterns. When the lateral centres vanish, we can only distinguish one-quarter of the previous patterns.

Another way to see the same result is to realize that the lack of lateral centres means our first cube has only two possibilities, instead of eight. Namely, it can be on the black side or on the white side. Either argument leads us to

$$\frac{(8!)^2}{4} \text{ or } 406\ 425\ 600 \text{ distinguishable patterns.}$$

The $2 \times 2 \times 2$ *small cube* has eight corners, but no edges. But the corners follow the same principle as the $3 \times 3 \times 3$ cube. Any two can be exchanged, but no one cube can be twisted on its own. So it seems that there are

$$8! \times 3^7 \text{ or } 88\ 179\ 840$$

different patterns. But this is based on having visible centres. Now all six centres have vanished and so we cannot distinguish patterns which differ by any of the 24 rotations of the whole cube. (Alternatively, all placings of the first corner are equivalent, so there is just one possibility for it instead of $8 \times 3 = 24$.) Hence the number of distinguishable patterns is divided by 24, giving

$$3^6 \times 7! \text{ or } 3\ 674\ 160.$$

4.2 Group theory and the cube

By any turn or sequence of turns the cube goes from one state to another — it may even change and return to start (for example, by four quarter turns of a layer. The set of mathematical transformations of the cube must include the transformation of no change at all, called the identity transformation. Any transformation followed by another transformation (or, for that matter, by the same one again) is yet another transformation of the cube.

Any transformation can be undone by reversing the turns and the order of the turns.

These properties show that the transformations of the cube form what mathematicians call a *group*. The *elements* of our group are the transformations of the cube. Two different sequences of moves represent the same element or different elements according to whether they give the same pattern or different patterns when applied to start position.

Let us illustrate the concept of a group in another way. Consider a number line. The points on it will be our elements. A point can be transformed in various ways which change its position or leave it fixed. We can combine and undo these transformations. Hence, all possible changes of position along the number line form a group and the position corresponding to start on the cube is the starting point on the number line marked by 0.

Just as a transformation of the cube can be characterized by the pattern reached from start, whatever the sequence of moves which has brought it about, so a transformation along the number line can be characterized by the point reached from the starting point.

The following list of similarities and differences can be extended.

Start	0
Make some turns from start	Move from 0
Apply a sequence of moves to start, then another sequence of moves	Make a movement from 0, then another movement
From start, apply the same two sequences, carried out in the reverse order. (The resulting pattern may or may not be the same)	Make the same movements from 0, but in reverse order. (The same point is reached)

Though the elements of both groups are *changes* from one pattern or point to another, these changes can be identified by the *states* reached when applied to a particular state (start or 0). The group of the cube has as many elements as there are cube patterns (4.3×10^{19}); the points on the number line are infinite in number, even if only integers are considered.

However, the concept of a group is wider than these two examples suggest. A group always includes an *operation* which combines two elements of the set to form a third. In each of our examples, the elements are transformations and the operation is: 'is followed by'. This operation is rather natural and hence perhaps somewhat hidden from view. Let us look more closely at an example on the number line.

On a number line a movement from point $+3$ to point -2, in any manner, can be characterized by -5, the point which would be reached by travelling the same way starting from 0 (Figure 4.8).

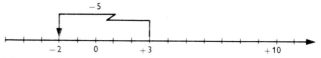

Figure 4.8　Any move from $+3$ to -2 is a change of position which can be characterized by -5

This is the first operation. The second consists of going from -2 to $+7$, again in any manner. Starting from 0, the point 9 would be reached, therefore this operation can be represented by $+9$ (Figure 4.9). The two movements together produce a movement from $+3$ to

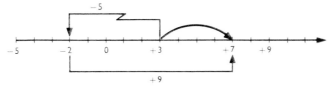

Figure 4.9 The change of position from -2 to $+7$ is $+9$. From $+3$ to $+7$ it is $+4$

$+7$. Starting from 0 this movement would bring us to the point $+4$. The two operations -5 and $+9$ therefore combine to give the change $+4$.

This is the same as adding vectors along a line.

For a third example of a group, consider stretchings and shrinkings applied to a vector. The role of the point 0 is now taken by a vector from 0 to 1, as at the top of Figure 4.10. A sixfold stretching gives the second vector in Figure 4.10 and a one-third-fold stretching (actually a shrinking) combine to give a $6 \times \frac{1}{3}$ or a twofold stretching shown in the last vector of Figure 4.10. These changes involve multiplication rather than addition. The most widely accepted notations for the cube (and groups in general) are multiplicative, which implies power notation for equal factors. This convention is partly due to the fact that the operation of this group is not commutative — the results of two sequences may be different if carried out in reverse order, and the multiplication of certain entities, for example, matrices, is not commutative, while addition is always taken as commutative.

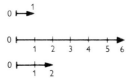

Figure 4.10 Sixfold stretching followed by a one-third 'stretching' (or threefold) shrinking)

The group of the whole cube is too vast to be surveyed; only some of the many subgroups can be considered. We shall restrict our set of turns in some ways and consider the subgroup of all patterns which can be achieved using our limited set of turns.

a) *Half turns of two adjacent faces.* This has been explored in section 2.2; only 12 patterns are possible. These are shown in Figure 2.13 (beginning from start).

The maximum number of moves required to restore the cube from any of these states is six, which is the maximum number of moves required to reach one state from another state. Although this number is known as the *diameter* of the group, the geometrical image here suggests a semiperimeter.

(b) *Half turns of three faces adjacent to a corner.* In this case there are three possible moves, F^2, T^2, R^2 (*fi, ti, ri*); from start, these produce the three different states shown in Figure 4.11. From each

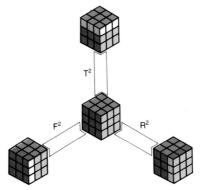

Figure 4.11 Half turns of three adjacent faces: three states can be reached in one move

state two new states can be reached, by turning either of the other two faces (Figure 4.12). This adds six new states. By a third move, 12 new states can be reached (Figure 4.13), and so on. The doubling of the number of states that can be reached at each move gives the numbers shown in Table 4.1.

A geometrical progression tends to infinity, yet the number of possible patterns is finite (calculated to be 2592). So the number of patterns cannot double indefinitely—we must get some repetition. For example, if the red–yellow alternating path is followed the same pattern is reached by making six moves one way or by six the other way (Figure 4.14). Different patterns occur in the process. Branches therefore find their way back instead of moving away indefinitely. Round trips similar to those of Figure 2.13 are formed along the red–blue and the blue–yellow alternating paths and other types of

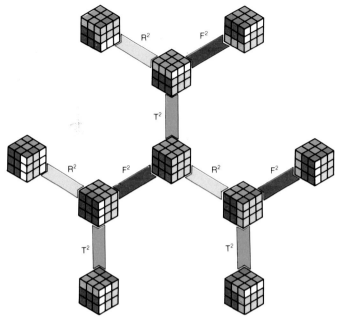

Figure 4.12 Half turns of three adjacent faces: six more states can be reached at the second move

round trips occur on other paths. Thus a number in Table 4.1 is the maximum number of patterns which can occur. Only computers can calculate how far away start is from the farthest state. But since there are 2592 possible patterns, it can be deduced from Table 4.1 that there must be states *more than nine moves away* from start.

Table 4.1 Number of possible patterns which can be reached by half turns of three layers adjacent to a corner

	Number of patterns: 2592						
Moves	0	1	2	3	...	9	10
Possible new patterns by last move	1	3	6	12	...	768	1536
Possible patterns up to last move	1	4	10	22	...	1534	3070

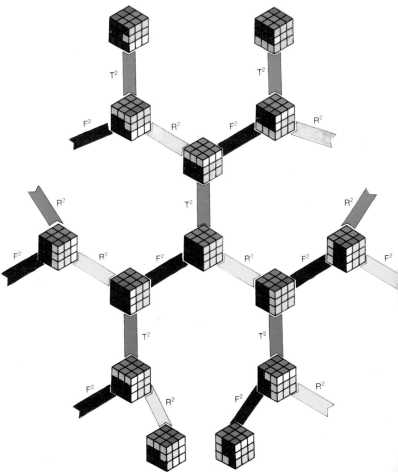

Figure 4.13   Half turns of three adjacent faces: 12 new states are reached at the third move

Figure 4.14   Two identical states reached in two different ways

(c) *Half turns of middle layers.* The previous subgroup had three possible moves from each state. We look at another subgroup with this property. We permit only the moves F_s^2, T_s^2, R_s^2.

As far as patterns are concerned these are the same as the *middle-layer half turns* of F_m^2 (*fim*), T_m^2 (*tim*) or R_m^2 (*rim*). (Moving the cube as a whole, or viewing it from different sides, does not affect the patterns.) Each of our half turns should be counted as one move. A map of this subgroup starts in the same way as the previous subgroup (see Figure 4.11). Yet at the next move matchings occur:

$$F_m^2 T_m^2 = T_m^2 F_m^2, \; F_m^2 R_m^2 = R_m^2 F_m^2, \; T_m^2 R_m^2 = R_m^2 T_m^2.$$

And after a half turn of each of the three middle layers the pattern is always the same whatever the order of the three half turns ($F_m^2 T_m^2 R_m^2 = F_m^2 R_m^2 T_m^2 = \ldots$). This pattern is known as the *pons asinorum* [or the $6 \times -DS$]. The present subgroup turns out to have only eight elements and its map has a simple structure, that of the edges of a cube (Figure 4.15 and Table 4.2). Its diameter (the

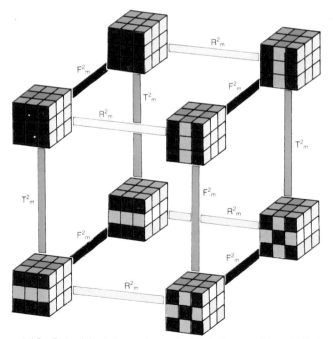

Figure 4.15 Only eight states can be reached by half turns of the middle layers

Table 4.2 Number of patterns which can be reached by half turns of middle layers

	Number of patterns: 8			
Moves	0	1	2	3
New patterns by last move	1	3	3	1
Patterns up to last move	1	4	7	8

maximum number of moves required to reach a pattern) is three and there is only one state three moves away. There is also only one state three moves away from any state. Maps of groups always display this kind of symmetry.

(d) *Half turns of any faces.* Now restrict the turns to half turns of any of the six faces. Then, at any point of the map, six choices are open, including the reversal of the last move. This would suggest that the map starts as in Figure 4.16 and Figure 4.17. Geometrical progression gives Table 4.3 for the numbers of possible patterns. The number of different patterns, 663 552 is given without proof. A breakdown of the geometric progression is unavoidable.

In fact the breakdown soon appears, with matchings occurring on the second move, since turning two opposite faces always produces the same pattern independently of the order of the moves. Hence

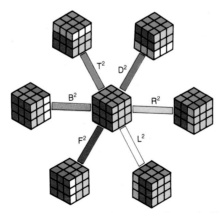

Figure 4.16 Half turns of faces: six states can be reached in one move

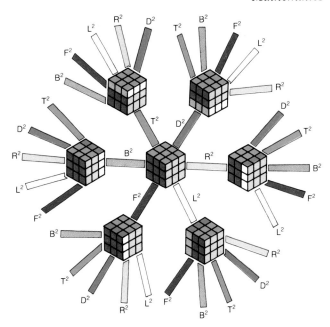

Figure 4.17 Half turns of faces: 6 × 5 new states at the second move?

Table 4.3 Number of possible patterns which can be reached by half turns of any of the six faces

	Number of patterns: 663 552						
Moves	0	1	2	3	...	8	9
Possible new patterns by last move	1	6	30	150	...	468 750	2 343 750
Possible patterns up to last move	1	7	37	187	...	585 937	2 929 687

the continuation of Figure 4.16 is Figure 4.18 rather than Figure 4.17, and the numbers in the second column of Table 4.3 have to be corrected from 30 to 27 and from 37 to 34. With or without these corrections it follows that eight moves are not sufficient to reach every possible pattern. Hence the diameter of the group is nine or more.

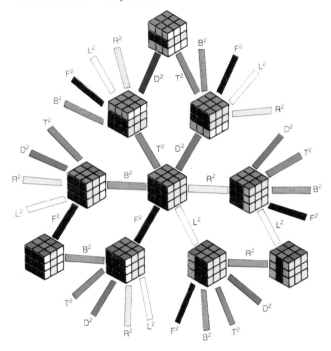

Figure 4.18   Half turns of faces: matchings reduce the number of new states

(e) *Any turns of any faces.* Now consider what happens when any turns are allowed. For the first move, there are three possible moves of each of the six faces leading to 18 patterns. Figure 4.19 shows the part of the network produced by one move.

In order to reach a new pattern at the second move, any of the five faces not affected by the first move should be turned. These five faces can each be moved in three possible ways and this gives 15 new patterns from each of the former 18 patterns, that is 18×15 which is decreased by 27, because of matchings. Figure 4.20 shows a third of the network produced by two moves. The nine matchings, indicated by ringed identical numbers, include $TD' = D'T$, $T'D' = D'T'$, $T'D^2 = D^2T'$. The numbers in Table 4.4 are calculated with similar matchings in mind. (The number of possible patterns was calculated in section 4.1.) The table shows that 17 moves are not sufficient to reach every possible pattern. More than half of them require at least 18 (see also section 2.12).

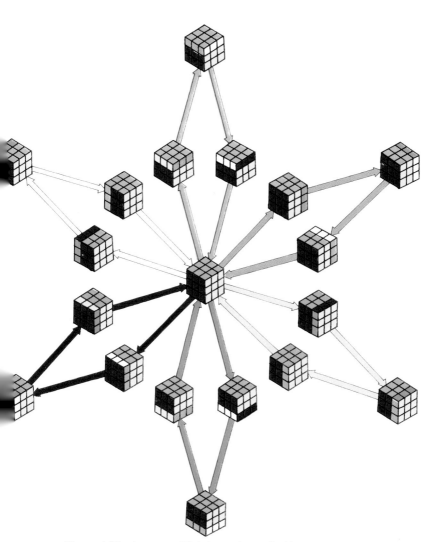

Figure 4.19 Any turns: 18 states can be reached in one move

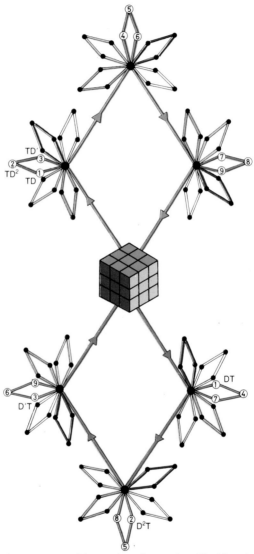

Figure 4.20   Any turns: matchings reduce the number 18×15 at the second move
by 27. One-third of them can be seen here

Table 4.4 Number of possible patterns reached when any turns are allowed

						Number of patterns: 4.3×10^{19}	
Moves	0	1	2	3	...	17	18
Possible new patterns by last move	1	18	243	3240	...	1.8×10^{19}	2.5×10^{20}
Possible patterns up to last move	1	19	262	3502	...	2.0×10^{19}	2.7×10^{20}

Table 4.5 Summary of the groups of the cube discussed in this chapter

Moves	Number of states (patterns)	Maximum number of moves required
Half turns of two adjacent faces	12	6
Half turns of three faces adjacent to a corner	2592	at least 10
Half turns of middle layers	8	3
Half turns of any faces	663 552	at least 9
Any turns of any faces	4.3×10^{19}	at least 18

5

THE UNIVERSE OF THE CUBE

György Marx

I respect the cube. I cannot fathom it. I do not want to learn how to do it from anybody else. Instead I want to experience the simple moves that hopelessly and mercilessly turn order into disorder. Whichever way I turn, disorder gives way to more disorder. It seems as hopeless to restore order as it is to get the spilt milk back into the jug or to mend a broken light bulb. The Magic Cube is a microcosm of the world and of nature. Our curiosity is excited by the cube's simplicity; we are challenged by its objectivity. You may fail to master the cube, but you will learn something about the depth of nature and the fascination of research from it.

At first glance, it seems that the cube consists of $3 \times 3 \times 3 = 27$ basic elements or small cubes. Though you cannot see the central small cube — the clue to the whole working mechanism is hidden — I like to think that a central small cube exists. The centres of the faces do not move in relation to each other. Only 20 of the 27 small cubes really move, and these have 48 coloured squares. There are 43 252 003 274 489 856 000 different ways of arranging the 20 pieces, that is, the 48 coloured squares. There are many possibilities, and there is only one ordered pattern. It is like looking for the proverbial needle in the haystack.

But is there really any difficulty? If you write the numbers 1 to 48 on 48 slips of paper, a young child can put them in the right order in minutes. Unfortunately, the little squares are somewhat different. Unlike chess, where the possible moves are restricted by rules which must be learnt, the Magic Cube has a hidden structure that, like the world itself, affects the moves we make and impose restrictions.

5.1 The universe of the cube

Place two Magic Cubes in front of you in their start positions (Figure 5.1). Choose a corner cube and remove its three coloured stickers. Then replace the stickers as they would look after a clockwise turn. The pattern (Figure 5.2) hardly differs from start. Now take this cube and try to match the state of the first cube by turns, in other words, try to unscramble it. Try as you will, you will not succeed. The twisted state, which you produced by rearranging the colours, cannot be achieved by turns. This state does not belong to the universe of the cube.

Figure 5.1 Start, the ordered state or the vacuum—a possible state

Figure 5.2 A clockwise twist (quark)—an impossible state

Now remove the three stickers and put them back so that they correspond to a turn of the corner cube in an anticlockwise direction (Figure 5.3). Again, it is impossible to get this position from start. Neither the pattern that resulted from a clockwise twist, nor the pattern that resulted from an anticlockwise twist belong to the world of the cube.

Figure 5.3 An anticlockwise twist (antiquark)—an impossible state

Now remove the three facelets of the two opposite corners of the front face and put them back as if these corners had both been turned clockwise (Figure 5.4). You have made a pattern containing two twists, which is also unattainable by turns. The removing and replacing of the colours is a *deus ex machina* defying the laws of the cube, and can only be used to play a joke. If you manipulate the cube by taking off the colours and sticking them back on differently, order cannot be restored.

Figure 5.4 Two clockwise twists (two quarks)—an impossible state

Now take off the three colours of three corner cubes and put them back as if the three small cubes had all been turned clockwise (Figure 5.5). Astonishing as it may be, you can get to this pattern from start—it belongs to the universe of the cube. You can also get the pattern of a clockwise and an anticlockwise twist (Figure 5.6). Such unexpected features are the fascination of the cube—and of the natural world too.

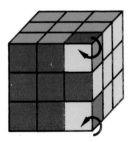

Figure 5.5 Three clockwise twists (three quarks)—a possible state

Figure 5.6 Twist and anticlockwise twist (meson)—a possible state

5.2 Quarks

We know the inside of both the atom and the nucleus. Science has split matter into infinitely many particles, but it is understood that the particles of the nucleus — the proton and the neutron — consist of three basic particles, known as quarks.[1] In 1969 it was discovered that other particles, known as mesons which are produced in the collision of nuclei, contain both a quark and an antiquark. This discovery was awarded a Nobel prize and attempts were made to isolate the quark — that is, to produce an independent free quark. Money and manpower were not spared, but this basic building block was never found. It is now accepted as a law of nature that nuclear particles are built up of quarks and that natural particles consist of three quarks, three antiquarks, one quark and one antiquark or combinations of these two types. In building, we produce standard bricks and lay them individually, but nature has a more interesting method. She never produces one quark-brick or combines two identical ones, but only combines *three* identical ones or *two* opposite ones.

The ways of nature are not entirely comprehensible. My students ask me what the quark is, whether it is the hidden brick of nature, or her archetype, or the archetype of the human mind attempting to get its bearings on reality, or a figment of our imagination. Solomon W. Golomb[2] was the first to point out that the cube is a model of the quark-structure of nature. He said that Rubik's cube was a form of mathematical and physical reality and that quarks, the constituents of particles, are another form of the same reality. He pointed out that although neither of them is a perfect model of the other, they are both described by a symmetrical group of permitted transformations, and that mathematicians are happy to recognize many unexpected similarities between them and encouraged to find further connections between the symmetries of the two phenomena. I consider the best way I can answer my students' questions is to show them a completed cube (which is a model of the vacuum state of maximal symmetry) and a cube with three twisted corners (see Figure 5.5), and ask them what changes they observe on the second cube and whether they can see the three quarks. They look at it and notice the twisted three corners. Then I ask them to produce by turns, a single twisted cube — a single change from the start or vacuum. As they cannot do this, they can then imagine how our universe never produces a free quark, although it is full of quark triples.

5.3   Conservation laws

Why is it impossible to attain the 'free quark state' (see Figure 5.2) from the vacuum? The answer must be in the nature of the turns permitted by the structure of the cube. We must look for the cause of this restriction in mathematics.

Take a cube at start and mark the corner facelets on the front red face and the back orange face with a zero. This means one facelet of each corner is marked. Write $+\frac{1}{3}$ and $-\frac{1}{3}$ (or simply a $+$ and a $-$ sign) on the other corner facelets, so that signs 0, $+$ and $-$ at each corner are in anticlockwise order. The signs can be glued or written on the facelets (Figure 5.7.) (This is equivalent to the marking in Figure 4.4 if one replaces $\frac{1}{3}$, $\frac{2}{3}$, by $-\frac{1}{3}$, $+\frac{1}{3}$ respectively.)

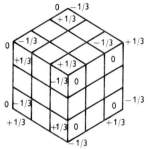

Figure 5.7   Adding the numbers on the front and back faces gives $B=0$

Add together the numbers on the front face and those on the back face. You will get zero. Now make any turn and add the numbers on the front face (with a red centre) and those on the back face (with an orange centre) again. Again you get zero. Denote the sum of the numbers on the front and back faces by B which we call the *baryon number* of the state. After a turn, the number B sometimes increases or decreases by one and sometimes stays the same. It is a law of the turns that the baryon number B is always an integer.[3] Now let us consider the start or vacuum position (see Figures 5.1 and 5.7). Here $B=0$. Next consider the free quark state that is produced by rearranging the colours of the red–yellow–blue corner cube (Figures 5.2 and 5.8). Here $B=+\frac{1}{3}$. The initial value of B is zero, and throughout any sequence of turns, it always remains an integer. Thus, the free quark, with baryon number $B=+\frac{1}{3}$, can never be produced from the vacuum with number $B=0$. On the other hand, the quark triple (proton) with $B=1$ *can* be produced (see Figure

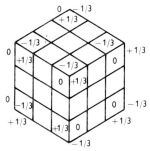

Figure 5.8   Adding the numbers on the front and back faces gives $B = +\frac{1}{3}$

5.5). It is the conservation of the integrality of the baryon numbers that makes the quark state unattainable.[4] All attainable states of the cube must have an integer baryon number. ('Another world' consisting of states with integer-plus-one-third baryon numbers, can be produced by taking the cube apart or by rearranging stickers.)

For a long time, man has tried to produce movement out of rest, to obtain energy out of a vacuum and to construct a perpetual motion. In the last century the principle of the conservation of energy was discovered and this led to further laws. The conservation of the baryon number[5] was discovered by Jenő (=Eugene P.) Wigner in 1949. This law states that $B = 0$ in a vacuum, and the constituent particles of the nuclei—protons and neutrons—have baryon numbers $B = +1$. If a proton consists of three identical quarks, the baryon number of one quark must be $+\frac{1}{3}$. Natural processes either conserve the baryon number or change it by an integer. A free quark, therefore, cannot be produced from a vacuum ($B = 0$) or from a proton ($B = 1$). The moves we can make on the Magic Cube help us understand this law.

5.4 Parity

Take two cubes at start. Remove the coloured facelets from one of the edge cubes and stick them back in the reverse order (Figure 5.9). This state cannot be achieved by turns, as it does not belong to the universe of the cube. However, if we reverse the colours of *two* edges (Figure 5.10), then we *can* obtain it from start—another hidden law of the cube which is a law of conservation. Six stripes on the six faces of the cube can be defined as follows (Figure 5.11):

Figure 5.9   A flipped edge — an impossible state

Figure 5.10   A double flip — a possible state

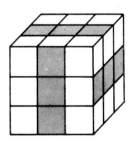

Figure 5.11   The stripes of the cube (stripes on opposite faces are parallel)

1. vertical bisecting stripes of the front and back faces (red and orange);

2. the right-to-left bisecting stripes of the top and down faces (blue and green);

3. the horizontal bisecting stripes of the left and right faces (white and yellow).

(These are the same as the frames of Figure 4.6.)

The stripes of opposite faces are parallel. Now write a + sign on the facelets which lie on these stripes, as shown in Figures 5.12 and 5.13. The total number of + signs is $N = 12$. If you turn the cube and count the number of + signs on the stripes, a conservation law concerning the possible values of N can be observed.[6]

Figure 5.12   The position of + signs at start

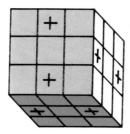

Figure 5.13   The position of + signs at start (rear view)

There are further laws governing the cube. Rearrange the cube to show the pattern in Figure 5.14. This can be done by removing the red–yellow–blue corner and the red–green–blue one and exchanging them. It cannot be done by turns, although the two conservation laws proved so far do not prohibit it since the values $B = 0$ and $N = 12$, are identical to those of start.[7]

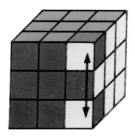

Figure 5.14   Another impossible state

However, making a thousand unsuccessful turns does not guarantee the impossibility of such a pattern — only mathematics can survey the cube's universe and patterns.

5.5   A survey of the cube's universe

The imaginary 'skeleton' of the cube is a spatial frame consisting of the invisible 'inside cube', and six immovable centre cubes. Around this skeleton, we have 20 cubes (eight corner cubes with three coloured facelets, and 12-edge cubes with two coloured facelets (Figure 5.15)). If we assemble these at random, there are over 500 million million million patterns. However, the cube has a structure that prohibits certain kind of moves — it is against the rules to take it apart, for instance. There are patterns that cannot be achieved.

Figure 5.15   Besides the imaginary inside cube there are six immovable centre cubes, eight corner cubes and 12 edge cubes

The mystery of the cube can be explored by the following moves. Take the cube in your left hand and turn the nearest layer a quarter turn (90°) in a clockwise direction. As in previous chapters, this move can be abbreviated as F (front). An F turn will transform the pattern of Figure 5.1 to the pattern shown in Figure 5.16. Two and three 90° turns of the same face give new patterns but four turns bring you back to the start. So we need only consider the moves F, F^2 and F^3 of the F face.

Figure 5.16   The cube after an F turn

Begin again at start and turn the right face of the cube by 90° (R). The pattern achieved is that of Figure 5.17. Turns R^2 and R^3 give different patterns and so do the turns of the top, left, back and down faces. But how many patterns can we get? One might think that the number of patterns equals the numbers of sequences of the form $F^aT^bL^cR^dB^eD^f$, where a, b, c, d, e, and f take on the values 1, 2, 3, 4. This would mean $4^6 = 4096$ different patterns, which does not seem too many. But then why is it that people get lost in the labyrinth of cube-turning?

Figure 5.17   The cube after an R turn

Take a cube at start and give the front face a quarter turn (Figure 5.16), and then make a quarter turn of the right face (Figure 5.18). Now take a second cube at start. Do an R turn first (Figure 5.17) and then an F turn (Figure 5.19) second. The same moves have been made, but in a different order. If you compare the pattern of the two cubes, you will see that they are quite different. This difference makes cube-turning a serious occupation. For two transformation F and R, if FR=RF, we say that F and R commute. If it holds for all of our transformations then our operation of combination is

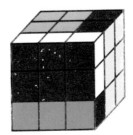

Figure 5.18   An F turn followed by an R turn

Figure 5.19   An R turn followed by an F turn

commutative. We have just seen that cube-turning is non-commutative. Because of this non-commutativity, the various sequences of our six basic moves yield 43 252 003 274 489 856 000 different patterns. This is less than the number mentioned above, 500 million million million. The conservation laws decrease the number of states possible, but do not make the cube simpler.[8] The non-commutativity of moves leads to an almost inexhaustible number of patterns, just as in chess.

But perhaps the cube puts chess somewhat in the shade.

We may find this non-commutativity somewhat alien, as we are used to 2×3 being equal to 3×2. Similarly, we may find working in three dimensions somewhat alien as we are used to working in two dimensions, for example, reading this sentence. But architects and sculptors work in three dimensions, such as Ernö Rubik and Terutoshi Ishige,[9] and those children who solve the cube instinctively. Modern mathematics (group theory) and modern physics (quantum theory) have discovered that a great many things do not commute. The shortest strategy for the cube has been worked out by Morwen Thistlethwaite, by the use of computers at the Polytechnic of the South Bank in London. The strategy restores the cube in 52 moves and it uses the complex structure of groups of turns. [52 has now been reduced to 50 — DS.]

Some quark combinations can be observed in high energy laboratories and others are missing. Physicists deduce the laws of matter and understand the structure of the universe from the observed states. Similarly we try to understand the inner structure of the cube from the observable colour patterns and the set of possible moves.

5.6   Disorder

If you make one turn a second, it will take you 1400 million million years to produce all 43 252 003 274 489 856 000 possibilities. During this time, there would be only one second when you, the tireless cube-turner, could cry 'I've done it!' The whole universe is only 14 thousand million years old.

Like masterpieces or computers, the cube cannot be mastered by chance. The cube, with its modest $3 \times 3 \times 3$ form illustrates the discovery of a hundred years ago that the world is built of a large number of atoms and the order of events cannot be reversed.

With the cube, however, we have a form of Ariadne's thread—we can write down all our turns, so we can get out of the labyrinth by making the turns in reverse.

Perfect order is one uniquely determined state. Disorder, on the other hand, is a sea of possibilities and can only be expressed by 20-digit numbers. If you make a 90° turn and you want to find the correct move that leads you back, then you have to select it from 12 possible moves. If you make two 90° turns, then you have 114 possibilities. After three 90° turns, there are 1068 alternatives, and after four 10 011. [Information supplied by Zoltán Kaufmann, student of physics.] We get hopelessly lost—disorder seems irrevocable, and the cube's chaos appears. The 12 choices at each step lead to total chaos—the inevitable result of physical laws.

As patterns succeed each other, they become more and more variegated. Physicists Éva Gajzágó, Péter Gnädig, and Ferenc Niedermayer have measured the cube's disorder by its variegation.

Let us take the cube's colours in the following order: white, blue, orange, red, yellow, and green. The variegation of the cube's patterns will be defined by numbers. For example, consider the rather disordered pattern shown by Figure 5.20: taking the face with a red centre, count the white, blue, orange, red, yellow and green squares. We get 2, 2, 0, 1, 1, 3. The sum of the squares of these numbers is $4+4+0+1+1+9=19$. The maximum value of this sum is 81, which occurs with the 9, 0, 0, 0, 0, 0 arrangement, that is, when the face is completely red.[10] The variegation of the face is defined as the difference of the maximum value and the actual value (in this case, $V_R=81-19=62$; when the face is one colour), $V_R=0$. The more variegated the face, the larger the difference. The most variegated arrangement occurs when all six colours appear on a face but none of them appears more than twice.

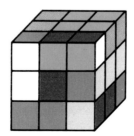

Figure 5.20 After five turns (F, R, T, L, D) the cube is variegated and difficult to unscramble

This means a 2, 2, 2, 1, 1, 1, distribution of colours; variegation in this case is 66.

To characterize the whole cube, we can determine the variegation of all the faces. In Figure 5.20 we have $V_R=62$ on the red face, $V_B=56$ on the blue, $V_O=62$ on the orange, $V_W=58$ on the white, $V_G=62$ on the green, and $V_Y=62$ on the yellow. The average variegation of the cube is defined as the arithmetic mean:

$$V=\tfrac{1}{6}(62+56+62+58+62+62)=60\tfrac{1}{3}.$$

Let us begin at start ($V=0$) and determine this average variegation as we make a sequence of random turns. Figure 5.21 shows the result of a test sequence of Éva Gajzágó. The average variegation approaches the maximum value after four turns and then oscillates near the maximum. Up to 30 turns it never gets back below 57.

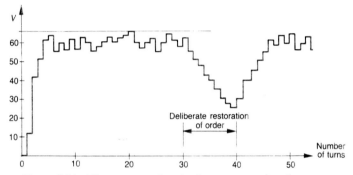

Figure 5.21   The average variegation in a sequence of random turns

As a test, the cube was taken over by a cubemaster after 30 turns. After ten moves, he had brought the average variegation down to 25, but the speed of decrease was regarded as slow. After a further ten moves, the cubemaster turned the cube at random. The variegation increased to 60 in a few moves.

The average variegation increases because there is only one pattern out of 43 252 003 274 489 856 000 which is perfectly ordered ($V=0$). There are 12 patterns that can be produced by one 90° turn from start and all of them have $V=24$. In 75 per cent of the cases, the value of V is greater than 30 after two turns. In other words, $V=0$ can be realized only by a single pattern while the larger values of V can be realized by more and more patterns. There are values of V that can be realized by trillions of patterns. If we turn the cube at random, then it is highly probable that we will get one of the many highly variegated patterns. There is only a very small probability that we will restore the cube.[11]

5.7 The science of the cube

The discoveries that every tangible object consists of many atoms, and that time is irreversible, as occurs in the mixing of gases and in the equalization of temperature, are interpreted in terms of disordered motion. If nothing was allowed to pass at a door between two rooms except oxygen from the left and nitrogen from the right, then the constituent parts of air would be divided. If nothing but fast-moving molecules were allowed to pass from the right and slow-moving molecules from the left, then one room would be heated by the cooling of the other. Maxwell used this idea to express the fact that the normally occurring increase of disorder is a statistical law—it is highly probable, but is not absolute. Its validity follows from the large number of possibilities. This idea is illustrated by the variegation of the cube.

We can ask ourselves how far final chaos is from perfect order, and how many turns we need to complete the cube from chaos. It is believed that we could find our way back to start in 20 or so moves, if we could select the perfect move at each step.

At the moment nobody knows the perfect algorithm, often called God's Algorithm. The chance of success is negligible, as it is highly probable that we would get lost in the labyrinth. Even grand masters of the cube need more than 50 turns to complete the cube after five random turns, and the well-known algorithms use several hundred moves.

We do not know how many atoms there are in the real world. Perhaps it is an infinite number and this is the cause of the large number of possible states, the direction of time, and the basic difference between geometry and reality. The cube with its 20 turning elements is a miniscule model of the world reflecting its temporal character. This is why it appeals to me.

If you want to restore the cube, you have to discover the nature of the turns and, step by step, the laws of its patterns. After some exploration you can devise a strategy that brings you back to start, but not efficiently. My student, Zoltán Kaufmann, devised a strategy in two weeks of logical work. Those who have not found the solution yet, and I am one of that unhappy number, have to resort to cheating—taking it apart or using glue to stick on new coloured facelets.

Bernie Greenberg, a computer expert at the Massachusetts Institute of Technology whose computer graphic of the cube was the cover illustration for the March 1981 issue of *Scientific American*, says, 'Cubism requires the would-be cubist to literally invent a science. Each solver must suggest areas of research to himself or herself, design experiments, find principles, build theories, reject them and so forth. It is the only puzzle that requires its solver to build a whole science.' An article in the same issue, Douglas R. Hofstadter, professor of computer science at MIT writes, 'Could Rubik and Ishige have dreamed that their invention would lead to a model and a metaphor for all that is profound and beautiful in science?'

Notes

1. 'Quark' is not a standard word in any language or literature. Its meaning in the German vernacular is identical to that of the Austrian *kvargli*, a figurative term for rubbish or good-for-nothing. Finnegan, the innkeeper in James Joyce's *Finnegan's Wake* is startled when drunk and cries out, 'Three quarks for Muster Mark!' This word was adopted by physicists as they were not certain whether or not quarks existed—and even if they do exist they cannot be observed.
2. A mathematician and electrical engineer at the University of Southern California.
3. See Chapter 4 for the proof.
4. More precisely: $3B$ is constant (mod 3).
5. *Barys* is a Greek word meaning heavy; the English word 'barometer' is derived from it.
6. See Chapter 4 for the formulation and proof of the second conservation law.
7 See Chapter 4.
8. The fact that B and N are always integers divides the number of states by 3 and 2. The inability to exchange a single pair of cubes halves the number of states. Hence only $\frac{1}{12}$ of all the possible patterns can be obtained.
9. In his monograph *Notes on Rubik's Magic Cube* (pp. 37–8) David Singmaster says; '... Terutoshi Ishige, a self-taught engineer and owner of a small ironworks about 100 km east of Tokyo, has also invented the Magic Cube. This was about five years ago, first in the $2 \times 2 \times 2$ form and then in the $3 \times 3 \times 3$ form with two somewhat different mechanisms .... The three Japanese patents ... show first dates of 1976 and 1977

second (issuing?) dates of 1978 and third (publication?) dates of 1980. Rubik's Hungarian patent shows three dates of 1975, 1976 and 1977, so he seems to have priority. Ishige's patent drawings show mechanisms conceptually equivalent to Rubik's, but quite different in details. To me, Ishige's designs appear less simple and less robust than Rubik's but I feel that further development would have led to something like Rubik's.'

10. In other words, the occurrence of colours on the cube's face can be characterized by a six-dimensional vector, the individual components of which give the numbers of occurrences of the corresponding colour. V_R, the variegation of the red face, is equal to the square of the length of this vector. Its maximum value is 81 (as at start), its minimum value is 15 (maximum variegation).

11. Physicists call the measure of disorder 'entropy'. It is a measure of the number of states which have the same appearance or behaviour. Random collisions impel particles into arrangements produced by more states and therefore the entropy of the universe increases. On the cube, if the number of different patterns with variegation V is an S-digit number, then S can be taken as the entropy that belongs to V. (More precisely, if there are $N(V)$ different patterns then the entropy of V can be defined as $S(V) = \log N(V)$, where the logarithm is to base 10.) The entropy of the cube is between 0 and 20 and cannot be larger, as the number of all cube-patterns is a 20-digit number. No exact results exist, but Gajzágó, Gnädig, and Niedermayer conjecture, that, with a few exceptions, increase of variegation implies increase in the number of patterns. The theory that random turns of the cube make it more variegated has yet to be proved.

6

MY FINGERS REMEMBER
The psychology of the cube

Tamás Vekerdi

When a radio reporter asked me about the Magic Cube I could only say that I didn't know the cube's secret but would very much like to.

On the other hand, Gábor Köves, a twelve-year-old schoolboy, says that after he has looked very carefully at each face of the scrambled cube and made a couple of turns or so, he can close his eyes and complete it.

A radio reporter asked him if he could *imagine* how the colours were moving.

'Yes, I know how it is moving, yes, I can *see* it, even though I've closed my eyes ... once in this direction, now twice downwards with this face, now turn this once in the direction in which I started, then the first face-upwards, then the top backwards twice, and backwards once — now it's finished.'

Listening to Gábor Köves I realized that he had scrambled the cube himself and remembered the turns visually and tactually. Try scambling the cube and working on it for days or weeks, and you will know the relief.

A translator friend of mine told me he had twiddled the Magic Cube for a week, and neglected his work and children. His wife added that the one week was really six weeks and that her husband had an idiotic look about him all the time. My friend pointed out that although logic or reason helped, he finally succeeded in cracking the mystery by turning the cube in different ways *and remembering how he had held it.*

This explanation is similar to that of Gábor Köves and is the most I have learned from anybody.

The reporter asked why children rather than outstanding mathematicians are the best cube-turners and how Gábor Köves could 'see' the colours with his eyes closed.

I would go further. Grown-ups as well as children can turn the cube with their eyes closed. Cubists who compete for championships say that *their hands remember.* They are not therefore relying exclusively on knowledge.

But why children? And what have the adult cubists got in common with children?

Some time ago, I was sceptical about the cube craze. A film director commented that this showed that I did not know the cube. She went on to make the interesting statement that when trying to solve the cube she had relived childhood experiences. She had seen the cube as a child sees the world *simultaneously from all sides.* I remembered children's drawings, which show everything the child knows to be there, whether or not he can see it.

He draws everything—he cannot see the garden through the wall, the clothes in the wardrobe, both legs of a rider or a room somehow spread out, but that is how he draws. There is no perspective or viewpoint—everything is seen *from all sides.* [So cubists see the cube as the cubist painters showed it—DS.]

If the cube were only a toy for mathematicians and for people who like mathematics, it would never have become an international success. So I am interested in trying to understand the essence of its fascination—that strange feeling, that characteristic way of looking at things. Could it be that the children and young people who can complete the cube quicker than mathematicians have a kind of photographic memory?

We know that children have a better visual memory. Many children and teenagers can remember a briefly shown picture for a longer time than adults can.

This can be proved by showing a picture to a group of children—a picture, say, of a market with people, stalls, and shops. In the background is a building with pillars and balconies. The picture is shown for 10 seconds. After a while you ask the children to tell you the number of apples in one of the stallkeepers' baskets.

They might say they have no idea, or try to guess, but one child might say, 'Just a moment,' and will close his eyes and begin to count.

He will recall the picture in his memory and count the apples!

He did not think of the number of apples when he was looking at the picture, and even if he had, *he would not have had time* to count them. But now he counts and tells you the exact number.

Next, you ask him how many pillars the building had, and again he remembers the picture, counts and gives you the exact number.

It is an astonishing experience but it is clear that children can 'develop' their memory images with the utmost precision. They have *eidetic* (Greek *eidos,* picture) *memories* (in other words, photographic) which start to fade at puberty, although we do not know why.

Children who think in pictures, somehow transform spatial relations and look at them from all sides simultaneously; they think concretely. In adulthood *abstract, conceptual* thinking takes over. *Words* and formulae are used and we are hemmed in by ordered spatial and temporal relations. It is characteristic of contemporary western civilization, that adults lose the ability to *create* and see *inner images.* However, our adult thinking means freedom — we are not limited by pictures and we can manipulate concepts and abstract notions such as algebra and mathematics quickly and efficiently. But this efficiency has its price — a healthy individual cannot entirely forego the creation of inner images and must — or at least should — be able to express stress, desires, and anguish in the form of daydreams. Creative work is based on an individual's private images and the ability to see things pictorially is encouraged in psychiatric treatment.

Perhaps then it would be true to say that the Magic Cube *brings out a kind of dormant ability* developing the spatial vision that was an inherent part of our childhood.

In *Scientific American* a cubist says, 'My fingers remember'. Is that possible?

We are rarely consciously aware of the sensation of movement — kinaesthesis — but we make use of it continuously; we perceive the states and movements of our body by visualizing them instantaneously. In a certain sense we imagine them.

The kinaesthetic sensation enables us to *feel* the form and size of an object, and to know what we are touching when we close our eyes. Our central nervous system, together with the other senses, constructs an *integral whole,* an *image,* from the instantaneous sensations.

Kinaesthesis is usually paired with *static* sense — the perception of our state of balance. These two sensations are called *proprio-ception,* or *self-sensation.* Contrary to the other senses, which tell us about the *outer world,* proprioception tell us about the *position of our body.* (Latin *proprius,* own or proper.)

To a great extent proprioception is dependent on *imagined vision* and some psychologists are of the opinion that it is not a real sensation. They say that kinaesthetic and static stimuli never evoke independent perceptions but only visual pictures of memory or imagination. There is no doubt, however, that an inner visual perception exists, fingers touching and turning the cube really are 'seeing and remembering'.

Eidetic memory makes it possible to recall in memory the faces of the cube we cannot see at a particular moment—all six faces successively or simultaneously; kinaesthetic perception depends on the memory in our fingers.

In Sir Francis Galton's book *Enquiries into Human Faculty and its Development,* published in 1883, there is a chapter on internal vision—the faculty of being able to *see* with closed eyes and to recall a picture of an object precisely and deliberately. Galton discovered that some people could not understand his question—for them 'seeing' was a figure of speech. Galton found that a great number of people had this faculty of internal vision and some people, mainly women and children, found it quite natural that they could recall the most distinct *image* of, for example, their living-room, parents, and friends. They gave precise answers to questions about details and became irritated, if the investigator considered they were 'remembering' and not 'seeing'. Galton discovered speakers and pianists who visualized their speeches or music and strove to 'see it better' when they made mistakes. He also discovered people who could 'see' more in their internal vision than in reality. 'This kind of picture is clear and distinct' a painter says, 'but cannot be drawn'. Neither can the six faces of the cube.

This comparatively small group describe their experiences as a kind of *touching-view.* Their vision is not only visual but also tactile. Time and space are irrelevant; *different views are seen simultaneously.*

If we could speak to Galton's respondents they would perhaps say, 'If I take the cube in my hand, I can feel its six faces simultaneously with my fingers and my palms, and the mental vision gives a similar type of feeling'.

Of course, to produce this feeling, the cube must not be too large, or it will not fit into our hand. Neither must it be too small, or it will be lost.

The Magic Cube seems to be an ideal size—it is made to measure. I have seen, touched and turned some imitations of the Magic Cube that were somewhat smaller or somewhat larger and none of them can be compared to the characteristic touch of the Magic Cube.

It is a *sensory experience* to touch, grasp, and turn the Magic Cube. Its surface is pleasing and it neither slips nor sticks. It has a stable but somewhat elastic movement. It gives the impression of being well-constructed, it moves easily and is tough. I expected the wooden version to feel different, but, in a peculiar way, it gave rise to very similar sensations.

Sir Francis Galton discovered the ability of internal vision—the 'touching-view'—decades ago. Ernö Rubik, without intending to do so, gave us an object to elicit what Galton described.

I can imagine that the Magic Cube unconsciously triggers a special faculty of vision and touch in people who have such tendencies.

I would also hazard a somewhat bolder hypothesis. If we say that the easiest and quickest way of completing the cube is not calculation but the reliance on some sort of sensation in which the coordinates of space and time play no part, then it follows that a cubist reaches the border between consciousness and sub-consciousness—eidetic memory and internal vision, lack of spatial and temporal coordinates, are characteristics of this frontier of the mind. Visualizing an abstraction and experiencing inner perception are a form of meditative exercise. Maybe it is not going too far to assert that fanatic cubists attain some kind of mechanically produced meditative state.

In modern psychology, archetypes, first used by C. G. Jung, denote physical fields of force embodying the accumulated experiences of mankind in the collective unconsciousness. These fields affect our way of life, our response to certain situations, our mental attitudes, actions, and choices, and rise into the consciousness in the form of certain signs, symbols, and figures which are constant or even identical in people of different cultures and at different times and places. Archetypes recur constantly, in rites, ceremonies, cult objects, the arts, dreams, visions, and revelations, and even in science. The cross, circle, triangle, square are all archetypal symbols. The number three has archetypal connections: there are divine trinities in several religions. The figure four is embodied in the four cardinal points of the four seasons. The sphere, pyramid and cube are also archetypal symbols; the world-process is described as a disintegration of the ordered divine cosmos into chaos and a struggle back to order and harmony both in Hindu and Judaic–Christian mythologies and in Hegelianism.

It is perhaps not overstating the case to say that the cube is an archetype. Cubic in form, it has square faces and leads from order to chaos and then, with great effort, back to the seemingly irretrievable order.

In archetypal conceptions chaos is paired with *freedom*. The number of possible states of the Magic Cube is about 4×10^{19}. However, the number of effective solutions, leading back to the ordered cosmos, is *limited*. There are millions of different possible ways, but there is *only one correct final state*.

It is an archetypal situation that an individual, in the course of his life sometimes has to destroy everything he has achieved. Jung said that the greater good is always an enemy of the good. The personal evolution of the individual, the so-called individualization process, requires a clean sweep time after time. Jung adds, that this process is accompanied by a natural fear, as we can never be sure whether we can rebuild but, on the other hand, the individualization process urges us on and if we resist we might fall ill. However, this urge can also be misleading, as it can be destructive.

The cube invites us to destroy what we have achieved. If we dare to do so, our actions may prove constructive and we may get a better result. But also present is the fear that we will never get so far again if we give up the results achieved so far.

6.1 The cube is a game

Both card games and football have cultural origins. Football, or at least the ball-game that was its predecessor, was played by the priests of the old Indian cultures, and playing cards originated in Egypt. [China seems to be the most generally accepted origin of both football and cards—DS.] In those days people wanted either to evoke the favour of the gods or to foresee the future. The ancient principle of a game on the one hand demands *skill and ability*, and on the other contains *a large built-in element of chance*. Games imitate the situation of man in the world, which, as with a game, he can get to know and to a great extent, master. However, it always turns out that he knows less than he thought he did.

The Magic Cube also satisfies the ancient principle of a game.

The heroes of stories and myths frequently have to solve riddles and puzzles, or find their way out of a labyrinth, or a large forest, where gods, demigods (and other fabulous creatures) live. In Egypt, the Sphinx watches over the pyramids (with their square bases and triangular sides); in Greece, she devastated Thebes, devouring the citizens, until her riddle was finally solved by Oedipus.

'Man is a real man when he is playing,' Schiller says. In the arts and sciences we play with questions and strive to find the answers, and by doing so acquire knowledge.

The mathematicians and physicists, who are cubists assert that many new questions arise while one is turning the cube.

However, I would not call them the cube experts, I prefer to give the title to the people who turn the cube in a completely self-preoccupied way, and perhaps do not even look at it but let it take them to forgotten areas of the human mind where it awakens dormant childhood abilities.

In Ernö Rubik's workshop I saw a figure of a man; its body is a Magic Cube, with legs and hands and a head sticking out of the cubes. This man enclosed in the cube resembled a man in the stocks. But although he is in the stocks he has incredible freedom of movement. When I first saw this statue I felt something of the magic of the cube in this sculpture—great freedom, but also great limitations.

I admire cubists. Take the following example.

We take the tube at the East Station, and a young man is sitting and staring into space. He is not even looking at the Magic Cube in his hand, but is vigorously turning it.

My travelling companion, whose family has worked with the cube for years but can only complete one face and the edges, wants to tell him to be careful as he will not be able to complete it.

The cubist is in another world. Why is he bothering to turn it if he is not even looking at it?

We arrive at our destination.

He stands up, and holds the Magic Cube in his hand.

All six faces are completed.

And he did not even look at it.

7

AFTERWORD

David Singmaster

Several years have passed since this book was written in Hungarian. This is really only apparent in that Rubik's biographical and historical account leaves off in 1981. It seems appropriate to add an afterword giving the history of the Magic Cube since that time.

In March 1981, Douglas Hofstadter wrote an exciting article on the Magic Cube in *Scientific American*, with a cover picture. This started the main phase of cube mania. The cube craze lasted a remarkably long time, starting in late 1980 and lasting on into 1983, longer than any other puzzle craze, except for the crossword fad of the 1920s. During this period, the cube inspired hundreds of new puzzles, ranging from straightforward size and pattern variations, through variations of shape using adaptations of Rubik's mechanism, to puzzles which were only related via the underlying mathematical theory. There were Rubik Cube clubs, tee-shirts, several newsletters, jigsaws, jewellery, a World Championship in 1982, a commemorative postage stamp, further awards, even piracy and lawsuits. I will describe some of all this below.

7.1 The Magic Cube since 1981

The Magic Cube won the UK Toy of the Year again in 1982 — an unprecedented event since no toy or game had ever won twice in a row before.

Rubik was awarded the Hungarian Order of Labour Gold Class in December 1981. He has now founded a Rubik Innovation Foundation to assist Hungarians in the 'practical realization of innovative ideas'.

Early in 1982, Ideal Toy introduced a 'Rubik's Range' of cube-related products. Some were devised by Rubik himself, others were selected by him. Later in the year, Ideal Toy was taken over by CBS and introduced the $4 \times 4 \times 4$ Rubik's Revenge. (The motivation for the name Revenge is unclear—Rubik says he has no desire for revenge on anyone!).

However, the great Magic Cube boom of 1981 encouraged unscrupulous manufacturers in Taiwan and South Korea to begin producing cheap 'knock-offs', generally known as 'pirate cubes'. These rapidly flooded the market in mid 1981, despite Ideal Toy's best legal efforts. Many shops were left with genuine Rubik's cubes that they had to sell at a loss.

Ideal Toy estimated sales of 10 million cubes in 1980 and they probably sold as many more in 1981. Politechnika told me that they sold over 10 million cubes in Hungary by mid-1982—that is more than the population of the country! It is estimated that the pirates sold several times as many as Ideal Toy, so it seems that perhaps 100 million cubes were sold in about three years. This amount and rate of sales far surpasses any similar product—Monopoly has sold about 90 million sets in its entire 50-year lifetime, Mastermind sold about 20 million in three years.

The cube craze tapered off in 1983 and shops were unwilling to buy any cube-like products as a result of the wholesale piracy, since when it has become almost impossible to find any for sale. Indeed CBS has remaindered its stocks. But I believe that we are starting to recover from this depressing state and that we will soon see Rubik's Magic Cube in its rightful place as one of the classic puzzles.

7.2   Rubik's offspring

In Chapter 1, Rubik describes several possible variations of the Magic Cube. Rubik himself, and many others, have devised several hundred variations and a remarkable number of these have actually been made. Further, many puzzles with very different mechanisms were produced, but they were clearly inspired by the concept of a moving-piece puzzle. I will concentrate on describing products which were actually made. In some cases, dozens of designs were produced by different people for the same object!

Starting at the trivial end, $1 \times 1 \times 1$ and $1 \times 1 \times 2$ cubes were made, and there was even a $1 \times 1 \times 3$ version. Less trivial was the $2 \times 3 \times 3$

Magic Domino, which Rubik invented and which has already been shown in Figure 1.8, number 6, and Figure 2.80.

The $4 \times 4 \times 4$ Rubik's Revenge was marketed in 1982–83. This had an interior sphere with grooves in which the centre-pieces could slide, and a simple but ingenious blocking to prevent the interior from becoming misaligned with the exterior. I do not know who devised the mechanism.

The $4 \times 4 \times 4$ turns out to be more complex than the ordinary Rubik's Magic Cube for two reasons. First, there are no fixed centres to give an orientation, so one must learn to keep the cube in a fixed orientation in space. Second, there is an additional parity problem which can leave you needing to exchange just two edge pieces. The number of patterns on the $4 \times 4 \times 4$ took me several tries to determine, but several others have confirmed the final result that there are 7 40119 68415 64901 86987 40939 74498 57433 60000 00000 different patterns when the faces are given solid colours. Several books appeared on the $4 \times 4 \times 4$ cube in 1982–83.

Nob Yoshigahara has devised a rather mind-boggling $4 \times 4 \times 4$. Start with several Rubik's Magic Cubes, take two apart and grind the pieces down to actual cubes. Then glue these small cubes onto the three faces which meet at a corner of a Rubik's Magic Cube and fill in along the edges and the corner. The results appears to be a $4 \times 4 \times 4$ cube, but the faces rotate eccentrically about the centres of the faces of the original Rubik's Magic Cube. This causes the cube's shape to change quite amazingly.

The $5 \times 5 \times 5$ cube was devised by Udo Krell and made in Hong Kong by Mèffert, of whom more later. It was only marketed in Germany (and perhaps Japan?), where it was called 'Rubik's Wahn', *Wahn* meaning illusion or delusion! It has a mechanism like a double-layered Rubik's Magic Cube, with the extra pieces worked in between the layers. Its solution does not introduce any new parity problems, but it obviously takes much longer to solve. However, it allows certain patterns that cannot be obtained on the $4 \times 4 \times 4$ cube, such as chessboard patterns on all six faces, so it is a good cube to have, and I am sad that it was not more widely available. It has 28287 09422 77741 85653 61803 33107 15032 82931 27731 98567 21347 21536 00000 00000 00000 patterns, when the faces are solidly coloured.

Having seen both the $4 \times 4 \times 4$ and $5 \times 5 \times 5$ cubes, a natural question is whether the $6 \times 6 \times 6$ is possible, and so on. When you turn a face, the corner piece moves out over the edge. For the

$6 \times 6 \times 6$, the corner continues to overlap the edge, but the overlap is so small as to be impracticable. For the $7 \times 7 \times 7$, there is no overlap at all and the corner-piece would fall off. One can make the inner layers of the cube thinner than the outer layers and this would permit making some of these larger cubes, but I do not know anyone who has done so. It is not hard to extend the theory of the $4 \times 4 \times 4$ to the general $n \times n \times n$, so perhaps there is no point in making such cubes. I have seen computer simulations of them.

The basic $3 \times 3 \times 3$ cube was subjected to hundreds of variations in size, pattern, and shape.

7.2.1   Size variations

The smallest cubes I saw were 12.5 mm ($\frac{1}{2}$ in) on an edge and were used as earrings! One form did not turn at all, in other words it was a $1 \times 1 \times 1$, but another form turned on one axis, making it a $1 \times 1 \times 3$. The smallest real cubes were 19 mm ($\frac{3}{4}$ in) cubes, often on key chains or necklaces. Large display cubes were made, but the largest functioning ones were about double the standard size.

Incidentally, I wondered why the pieces in Rubik's Magic Cube have a 19 mm edge, making the whole cube have a 57 mm edge. Rubik said that he tried different sizes to determine which felt good to the hand. He tried both 18 mm and 20 mm pieces and the results did not feel as good as the 19 mm pieces.

7.2.2   Pattern variations

About half of the pattern variations simply replaced the plain coloured stickers with patterned coloured stickers; for example, dice patterns, chessmen, fruit, flags, animals, card suits, etc. Generally this meant that a solution had also to orient the face-centre pieces, and this restores the factor of 4096 to the number of patterns, as mentioned in Chapter 1.

Two interesting versions are the following. First, take each sticker as divided into two triangles coloured black and white. Second (due to Alistair Shepherd), is to make all stickers have the same oriented pattern, such as a heart or an arrow or the letter A (Figure 7.1). Cubes were also adapted for the blind; the patterned cubes solved the problem of cubes for the colour-blind.

More interesting pattern variations had one picture spread over all nine stickers of one face of the cube. For example, there was a

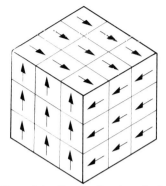

Figure 7.1 Shepherd's oriented cube

Royal Wedding Cube with portraits of Prince Charles and Lady Diana. You could mix their features and try to predict the appearance of their offspring! I also saw photos of nudes, advertisements for biscuits, animals, butterflies, and souvenir views of Venice and Hawaii. John White made a cube with the same pattern on all six faces (Figure 7.2). In fact all turns leave this cube in the same pattern!

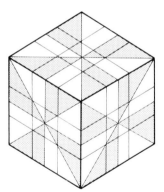

Figure 7.2 White's trivial cube

In a different direction, many people made their own colour variations by peeling and rearranging stickers from several cubes. For example, Tamás Varga and his colleagues made cubes with just two or three colours of stickers arranged in various ways on the faces. Politechnika made a Rubik's Mini Cube which was a 2×2×2 with just three colours, each colour placed on two adjacent faces.

This is more confusing than one first thinks—one can get to a point where one wants to twist a single corner. One can also remove the stickers from the edges or the corners or the centres or even from two of these. Removing just the centres gives a parity problem—one can get to a point where one wants to exchange two edges.

Other experimenters cut stickers and made cubes where the faces had two colours, splitting the face into two rectangles (Peter Redmill) or into two triangles (Neil J. Rubenking) (Figures 7.3 and 7.4).

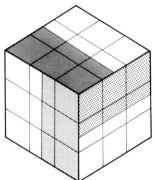 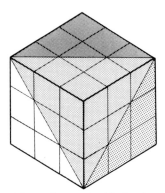

Figure 7.3 Redmill's cube   **Figure 7.4** Rubenking's cube

Rubik has described a pattern with four colours in Chapter 1 (Figure 1.7). The Italian distributors of Rubik's Magic Cube, Mondadori, provided a 12-coloured cube, where the colours were associated with the twelve edges (Figure 7.5). Mondadori also

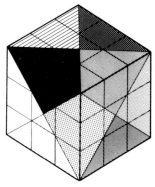

Figure 7.5   Mondadori's 12-colour cube. Each of the unshaded triangular region is a different colour, so that nine of the 12 colours are visible

produced a cube with magic squares on each face. In other versions of this, the numbers were printed on six colours of sticker, so the numbers were actually irrelevant. The Mondadori version had all its stickers with a gold background, so you could not tell which number belonged on which face. Other Italian producers were remarkably inventive, producing cubes with the symbols of their political parties, the colours of their football teams and even with caricature faces of their party leaders.

Yet another direction of variation was the 'calendar cube' which had digits and letters arranged so that one could set the day, month and date on one face (Figure 7.6). This is more ingenious than it first sounds—there are just barely enough facelets of each type to accommodate the various digits and letters. This was devised by Marvin Silbermintz, of Ideal Toy in New York. I have seen American, English, German, Italian, French, and Japanese versions, though the Japanese one is just an adaptation of the American.

I have made a 'labyrinth cube' by taking a cube with all white stickers and drawing a continuous circuit through all 54 of the white facelets (Figure 7.7). This has never been restored by anyone!

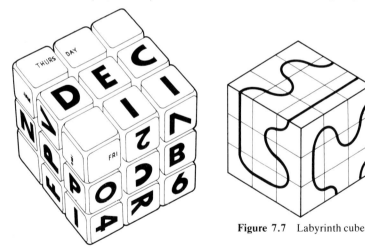

Figure 7.7   Labyrinth cube

Figure 7.6   Calendar cube

7.2.3   *Shape variations*

For the amateur in geometry, the variations of shape are the most fascinating. Rubik has systematically examined the possibilities and describes many of them in an article 'Magic polyhedrons' in his magazine, *Rubik's Logic and Fantasy in Space* (1983, No. 1, pp. 4–15). Douglas Hofstadter devoted an article to cubic variants in *Scientific American* (July 1982) and most of the variants considered by him are variants of shape. I will start with variants of the 3×3×3 and 2×2×2, then I will consider other shapes.

Beginning with the 3×3×3, the first variant to appear was the octagonal prism, formed by trimming four parallel edges from a cube (Figure 7.8). This is usually coloured with ten colours. It introduces a parity problem because some edge pieces can be flipped without showing. Also the shape varies as it is turned. I found a variation on this in Poland. Imagine the right face of Figure 7.8 rotated 90°. The resulting shape is given just two colours (Figure 7.9).

Now imagine our trimming carried out along two different axes. This produces a 'cushion'-shaped object, given 13 colours (Figure 7.10). A Polish version of this has the top and bottom face centres

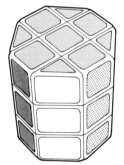

Figure 7.8   Octagonal prism

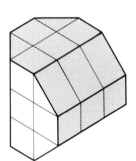

Figure 7.9   Polish prism

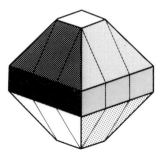

Figure 7.10   Cushion; the four sides around the middle and the eight sloping areas are given 12 colours; the top and bottom squares have a 13th colour

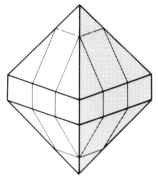

Figure 7.11 Polish cushion; a colour starts at the top peak and covers a slanting triangular area, then a side of the middle, then a lower slanting triangular area; this uses four colours

peaked to extend the sloping sides and uses just four colours (Figure 7.11).

If we repeat our edge trimming along the third axis, we obtain a truncated rhombic dodecahedron (Figure 7.12). If all the original face centres are peaked up to square pyramids extending the slopes, we get a genuine rhombic dodecahedron (Figure 7.13). These last two have 12 colours. I have only seen the latter in Poland. If, instead, we flatten out the corner pieces which our three

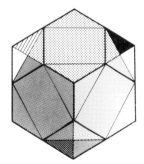

Figure 7.12 Truncated rhombic dodecahedron I

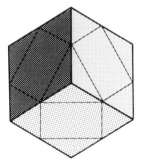

Figure 7.13 Rhombic dodecahedron; this is formed of 12 rhombuses, alternately meeting at three- and four-fold vertices; Figure 7.12 arises by cutting down the four-fold vertices; Figure 7.15 from cutting down the three-fold vertices

trimmings have left as triangular pyramids, we get the rhombi-cuboctahedron which is coloured with either three or ten colours (Figure 7.14). Yet another Polish version has flattened corners, but peaks the face centres into square pyramids, giving a peculiar shape which is a differently truncated rhombic dodecahedron and which is peculiarly six-coloured (see Figure 7.15).

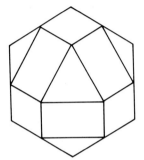

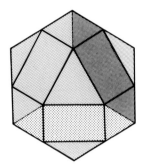

Figure 7.14 Rhombi-cuboctahed-ron; this arises from truncating the rhombic dodecahedron at both kinds of vertices

Figure 7.15 Truncated rhombic dodecahedron II; there are 12 different regions, but only six colours are used; the four similar regions going round the equator have the same colour; likewise for the four similar regions going from the right face over the top

As illustrated by the Polish prism, one does not need to have the trimmed edges all parallel. Nor does one need to trim four edges. One can easily build prisms with one, two or three parallel edges trimmed, and I have seen these on sale. I built a version with three perpendicular, but disjoint, edges trimmed (Figure 7.16).

Returning to the ordinary cube, we trim off the corners to obtain a truncated cube (Figure 7.17). The small triangles exposed at the corners are irrelevant and are coloured gold or silver.

If we make deeper cuts which pass through the midpoints of each edge of the cube, we obtain the cubo-octahedron or cuboctahedron, one of the semi-regular solids discovered by Archimedes (Figure 7.18). The exposed triangles are again irrelevant since the corner pieces retain small portions of the original cube face colours, but 14 colours were usually used for this.

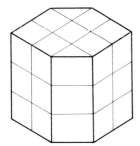

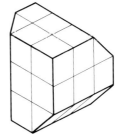

Figure 7.16　Two and three trimmed edges

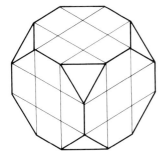

Figure 7.17　Truncated cube

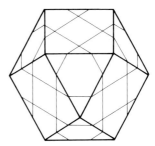

Figure 7.18　Cubo-octahedron

If one makes even deeper cuts which pass through the centres of each face of the cube, we get an octahedron (Figure 7.19). But the cutting plane now meets the corner piece in just one point, so it drops off the puzzle. No version of this was ever produced, but we will see an equivalent puzzle under octahedra. If we make the central layer of our $3 \times 3 \times 3$ thinner than the outer layers, we get face centres in an octahedron. This form was invented by Josef

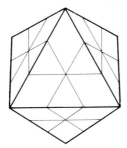

Figure 7.19　Cube truncated to an octahedron

Trajber (Figure 7.20). One can combine pieces from two or more of these variants to produce chimerical cubes. A unique variation looks like two cubes with a common edge (Figure 7.21). I first heard of this from Tony Fisher. About a year later, Ideal Toy marketed it in 1983 as Rubik's Mate, but Fisher called it Siamese cubes. In fact the two parts are independent and each behaves like a cube with the pieces along one edge glued together.

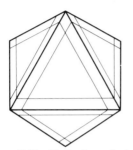

Figure 7.20   Trajber's octahedron

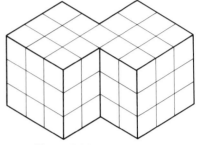

Figure 7.21   Siamese cubes

An easy variant of the cube is simply to round it out to a sphere. However, it is not so easy to make the sphere have six colours, though it was often done. Human faces also appeared, but a map of the world is the most interesting pattern for a sphere. A schematic map of the world was marketed as Rubik's World by Ideal Toy, and gave one the opportunity to sort out the world. The globes generally came with a semicircular holder which fitted into the poles and attached to a base (Figure 7.22).

East Germany has produced a series of four variants of the 2×2×2. The simplest is just an octahedron with each face having coloured dots at each corner, giving a kind of six-colouring, with

Figure 7.22   Rubik's world

Figure 7.23 Stella octangula

colours associated with corners rather than faces. If one puts tetrahedra on all the faces, the result is the stella octangula discovered by Kepler (Figure 7.23). One can view it either as two large interpenetrating tetrahedra, or as a large tetrahedron with small tetrahedra on the centre of each face. It is four-coloured in accordance with this last description — a large face and the small tetrahedron on it are one colour. If one puts tetrahedra on just four of the faces of the octahedron, the result is a large tetrahedron which rotates on its midplanes, that is, the planes which bisect four of the edges, cutting the tetrahedron in a square cross-section. This is naturally four-coloured. It changes shape as it turns and is thoroughly mystifying (Figure 7.24). It was invented by several people, including Rubik.

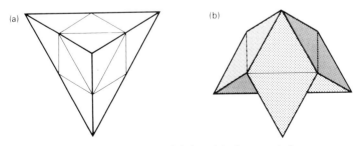

Figure 7.24 Tetrahedron 2 × 2 × 2, in its original state and after two moves

The last of the East German versions is again simple. Take an ordinary 2×2×2 and trim four parallel edges to leave a square prism (Figure 7.25). This also changes shape as you turn it, but it comes with just two colours—four pieces are solidly red and the others are solidly yellow. One of the standard shapes is the house shown and the puzzle is called the *Trick Haus.*

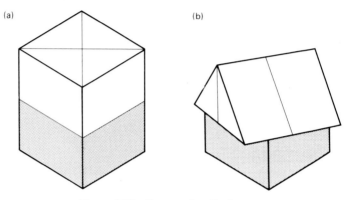

Figure 7.25   Square prism *Trick Haus*

7.2.4   *General remarks on other shapes*

The tetrahedron is the only regular solid whose faces are opposite to vertices, rather than to parallel faces. So the planes parallel to a face can also be viewed as planes cutting off corners. On the other regular polyhedra, these are distinct concepts. If the puzzle is made to turn on planes parallel to a face, we will say that its faces turn. If it turns on planes which cut off a corner, we will say that its corners turn. To illustrate this, Rubik's Magic Cube has *faces* that turn, while Trajber's octahedron has *corners* that turn.

Each polyhedron has a dual which is obtained by joining the centres of its adjacent faces. The dual of the tetrahedron is again a tetrahedron, but in the opposite orientation, like the two interlocked tetrahedra in the stella octangula. The other regular polyhedra occur in dual pairs: cube–octahedron; dodecahedron–icosahedron. The dual of the cubo-octahedron is the rhombic dodecahedron. Turning a face on a polyhedron is equivalent to turning a corner on its dual, so that the two kinds of turn are closely related, but the colourings of the polyhedron and of its dual are usually different so we cannot usually replace one kind of turn by its dual on the dual polyhedron. For example, Trajber's octahedron is the dual of Rubik's cube, but the cube corner corresponds to the octahedron face-centre and these pieces are differently coloured. Trajber's octahedron is equivalent to a cube with eight colours based on the corners, as in Figure 7.26.

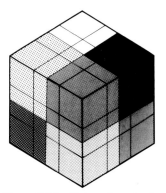

Figure 7.26 Eight-coloured cube

Besides the choice of turning faces or corners, the positioning of the planes of rotation gives quite different structures. This is most clearly seen with a triangular face with lines showing where it is cut by the rotation planes. Figure 7.27 shows the progression as the lines move in from the edges to the opposite corners. If there is a second line in each direction, we will expect them to be equally spaced as in Figure 7.28.

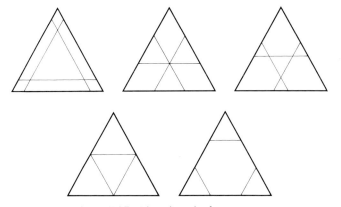

Figure 7.27 Five triangular face patterns

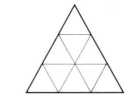

Figure 7.28   Triangle cut by six lines

With these general ideas, we see there is a large number of possible ways to imagine a Magic Polyhedron. I think all these ways have been contemplated by Rubik, and probably others by now. I will generally restrict myself to those which have actually been made. Rubik's mechanism is readily adapted to produce Magic Polyhedra with faces turning on nearby planes—one has only to modify the interior pieces to have the right number of axes and to modify the shapes of the pieces to correspond with the paths of rotation. Surprisingly, only the dodecahedral version was done in this simple way, as described below.

7.2.5   Tetrahedra

Uwe Mèffert, a Franco-German-Swiss engineer and inventor living in Hong Kong, made a tetrahedral puzzle in the early 1970s, but he did not see its commercial potential until the cube boom in 1981. He then marketed it as Pyraminx or Pyramix (Figure 7.29). It has two planes parallel to each face, so each face is cut into nine triangular facelets, as in Figure 7.28. It has three types of piece: four vertices; four corners, just under the vertices; and six edges. I have labelled these V, C, E in Figure 7.29. The vertices rotate independently of the rest of the puzzle and hence are trivial. The corners

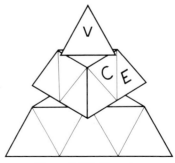

Figure 7.29   Pyraminx

are easily put right, so we have only the edges to worry about. Consequently, the Pyraminx is much simpler than the Magic Cube but it uses all the relevant mathematical ideas and so makes a good starting point for learning these ideas. Nicholas Hammond has shown that God's Algorithm is at most 21 moves (including the four trivial vertex moves).

[More recently, three people have found God's Algorithm. The maximal number of moves is 15 (including the four vertex moves).]

The tetrahedron also appeared in truncated forms; that is, without the vertex pieces and with the exposed parts of the corner either flat or rounded (Figure 7.30). The latter looks like a sphere cut by four tangent circles. Mèffert also considered having edges that turn, but no mechanism has been found for it. A stellated form of the Pyraminx with glittery stickers was a very attractive object, but only a few were made (Figure 7.31). Mèffert has obtained

(a) 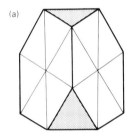 (b)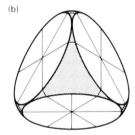

Figure 7.30   Truncated tetrahedron, two forms

Figure 7.31   Pyraminx star

designs for many different magic polyhedra and has produced several of them, as well as the $5 \times 5 \times 5$ cube.

7.2.6   Octahedra

Trajber's octahedron has already been described. Mèffert produced an octahedron with turning corners and faces looking like Figure 7.28 (Figure 7.32). This also appeared in a truncated form (Figure 7.33), which is equivalent to the object obtained by truncating a cube to an octahedron, as shown in Figure 7.19.

Rubik mentions an octahedron which turns on its midplanes, so it has turning faces and a face pattern like the fourth diagram in Figure 7.26, but this has not been produced (Figure 7.34).

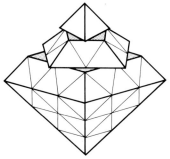
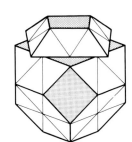

Figure 7.32   Pyraminx octahedron   **Figure 7.33**   Truncated octahedron

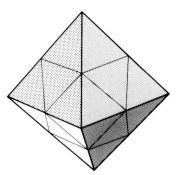

Figure 7.34   Rubik's octahedron

7.2.7   Cubes

Tony Durham, a London journalist, invented a cube with corners turning on the midplane which perpendicularly bisects the space diagonal of the cube. This plane bisects six of the edges of the cube and cuts the cube in a regular hexagon, as shown in Figure 7.35. There are four such planes and Durham calls this the four-axis mechanism. Mèffert produced a small number of these as Pyraminx Cube, but Hofstadter's second article named it the Skewb. Those who have one agree that it is one of the most intriguing cubic variants and hope that it will eventually be marketed. Durham also designed the dual shape to this, which is the octahedron invented by Rubik.

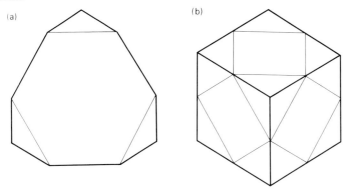

Figure 7.35 (a) Half of a Skewb; (b) Skewb

7.2.8. *Dodecahedra*

The Magic Dodecahedron was invented by several people. Two versions were manufactured, with slightly different face patterns. A Hungarian version was marketed in Europe and called Supernova in the United Kingdom. The other version was produced by Mèffert and marketed in the United States as Megaminx (Figure 7.36).

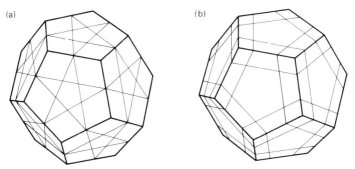

Figure 7.36 Magic Dodecahedron: (a) Supernova and (b) Megaminx

Tony Durham noted that one can locate a cube naturally in a dodecahedron, so his four-axis mechanism can be used. Now, if the faces of a Skewb are solidly coloured, the eight corners have recognizable orientation, but the six face centres do not. On the dual octahedron, with solidly coloured faces, the six corners have recognizable orientation but the eight face centres do not. When the four-axis mechanism is embedded in a dodecahedron with solidly coloured faces, all 14 pieces have recognizable orientation. This effect could be obtained on the Skewb or the dual octahedron by using patterns such that the face-centre orientations can be recognized.

Durham also proposed a dodecahedron with planes set further in, giving two parallel planes of rotation between each pair of opposite faces, while Rubik has wondered if one can have just one plane of rotation halfway between opposite faces. Durham has a design for his, but Rubik indicated that he has not found a mechanism for his proposal.

Besides the five well-known regular polyhedra, there are four stellated regular polyhedra which were discovered by Kepler and Poinsot. One of these, the Great Dodecahedron, was made to have rotating faces by Adam Alexander and was marketed by Ideal Toy as Alexander's Star. Conceptually, it is the same as the edges of a Magic Dodecahedron, but Alexander uses only six colours—opposite faces are coloured the same—and this gives a parity problem in solving it (Figure 7.37).

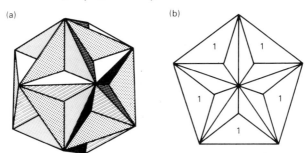

Figure 7.37 (a) Great Dodecahedron and (b) one face of it; in Alexander's Star, the regions marked 1 on the face are all the same colour

7.2.9 Icosahedra

For the icosahedron, there is a natural way to turn corners. The five triangular faces at a vertex form a pentagonal pyramid which can be turned on its base plane. This is unique among our polyhedra in that the faces are not subdivided at all—the whole faces are the pieces of the puzzle. This is actually equivalent to just the vertices of a Magic Dodecahedron. Only one icosahedral puzzle was made, called the Impossi-Ball, produced by Mèffert and marketed in the United States (Figure 7.38). This was an icosahedron rounded out to spherical shape. The pieces are spherical triangles and the five at a vertex no longer sit on a plane. Consequently a novel but simple mechanism was devised which allows an up and down motion while rotating.

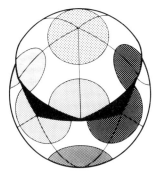

Figure 7.38 Impossi-Ball in mid-turn

7.3 Rubik's relatives

Many puzzles with moveable pieces have been inspired by Rubik's Magic Cube, even if they have very different mechanisms. Quite a number were actually produced. I shall describe these briefly.

The simplest types are related to Sam Loyd's 15 Puzzle (see Figure 1.6). There were several forms where the rows and columns were moved on plungers, which either obstructed some of the moves or required supplementary positions of a different nature. The three-dimensional sliding cube puzzle, which I mention in the Foreword, was produced in several forms (Figure 7.39). The most popular way of three-dimensionalizing the 15 Puzzle was to wrap it around a vertical cylinder consisting of several discs on an axis. The pieces could then slide vertically into the blank space and each row was on a disc which could rotate about the axis. Some versions of these puzzles had plungers or discs locked together to reduce the possible moves. One of the cylindrical puzzles had vertical plungers

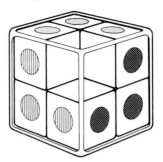

Figure 7.39   Sliding cube puzzle

instead of a blank space. These could also be flattened out into circular rings. The presence of a blank space or supplementary space(s) (for the plunger(s)) means that the underlying group theory is somewhat obscured, but it still provides the techniques for determining the possible patterns and algorithms for achieving them.

Closer to Rubik's Magic Cube are 'interlocking cycle' puzzles where several rings of pieces cross each other. Endre Pap, a Hungarian engineer, invented a flat version with two rings which was marketed as the Hungarian Rings (Figure 7.40). The idea was not entirely new, as there is an 1893 patent for it. Another Hungarian product, called Equator, had three rings forming three orthogonal great circles on a sphere (Figure 7.41). It came in a four-colour pattern and with a world map. Other flat puzzles had rotating interlocking discs. This was invented by several people —

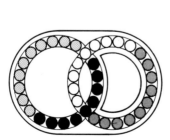

Figure 7.40   Hungarian Rings

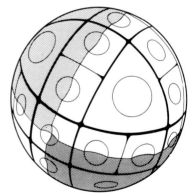

Figure 7.41   Equator

the version shown was actually made in Israel from the design of Dror Rom and is called Rom's Rings (Figure 7.42). Raoul Raba devised a more complex disc puzzle where the pieces were not all the same and a turn could block other turns. This was made by Pentangle in the United Kingdom as Rotascope (Figure 7.43).

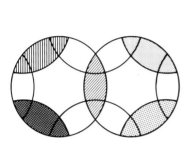

Figure 7.42 Rom's Rings

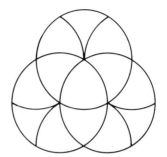

Figure 7.43 Raba's Rotascope. This comes with a figure on the pieces

A further type of closely related puzzle is what I call the 'switchable cycle' puzzles. Here one has several disjoint cycles of pieces with some mechanism for splitting the cycles and rejoining the parts into other cycles. The best known example was the Orb-It (UK) or Orb (US), which had four parallel tracks on a sphere (Figure 7.44). One hemisphere could be turned to eight positions with respect to the other, causing the half-tracks to join up in several different ways. This was marketed in both Europe and the United States. Another version, called VIP-Sphere or Logi-Sphere, allowed the hemispheres to have just two relative positions. It was marketed in Germany. A Hungarian Puck, due to Andràs Vègh, had a rotating circle and a surrounding ring which could turn 180° about a diameter (Figure 7.45).

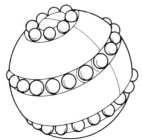

Figure 7.44 Orb-It

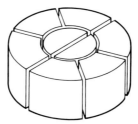

Figure 7.45   Puck. This comes with a picture on the pieces

There can be no greater proof of the fertility of Rubik's invention than the remarkable number of progeny that it has produced.

7.4   Magic Cube theory

The Magic Cube continues to fascinate mathematicians and computer scientists and considerable work has been done since 1981. This is not the place to discuss this work in detail, but a few points may be worth relating.

The $2 \times 2 \times 2$ Magic Cube has been completely resolved — the maximum number of moves required to restore it is 11 and there are 2644 patterns which require this maximum.

The closest approach to God's Algorithm on the $3 \times 3 \times 3$ Magic Cube is due to Morwen Thistlethwaite, who uses just four stages, requiring 52 moves. The last of his stages uses the group generated by all the 180° turns of faces, such as R^2, F^2, ... Thistlethwaite could solve this group in 17 moves, but conjectured that 15 was enough. This group has since been completely analysed and 15 moves is indeed sufficient, so Thistlethwaite's method is reduced to at most 50 moves. I have looked at several counting arguments and the best show that some positions require at least 21 90° turns or 18 face turns (which can be either 90° or 180° turns).

The $n \times n \times n$ cube has been studied by several people and the number of patterns has been found, but the formula is not simple. The $3 \times 3 \times 3 \ldots \times 3$ (n times) cube has also been solved.

It is possible to study magic polyhedra in general. One can even take any planar graph and imagine its faces able to rotate. I have found the possible patterns on any such graph and have shown that the careful use of commutators allows one to achieve all the possible patterns. Interestingly, the hardest versions are those with the fewest faces. Indeed, this has been found to be a characteristic of all moving piece puzzles—the more different moves one can make, the easier it is solve! [Wolfgang Hintze points out that face turns on an octahedron or icosahedron give rise to a new type of triangular piece—see Figure 7.46— and he has found the theory for such polyhedra.]

Figure 7.46   An octahedron with turning faces

7.5   Championships

A World Championship was held in Budapest on 5 June 1982; 19 national contests produced champions who came to Budapest. The US Champion, Minh Thai, won with a time of 22.95 seconds. This, like most competition results, is slower than one expects because new cubes are used which are not 'broken in'. In practice with used cubes, times of 17 to 20 seconds were being recorded. The US Championship was shown on the American TV programme 'That's Incredible' and the World Championship was widely covered by all media. Hungary issued a special stamp and cover to commemorate the day.

There was a second US Championship in 1983 and Minh Thai easily retained his title.

I have only heard of one Championship for Rubik's Revenge. This was in the United Kingdom on 25 June 1983 and the UK Rubik's Cube Champion, Julian Chilvers, won in 119 seconds.

7.6   Newsletters and magazines

The earliest newsletter was my own *Cubic Circular:* No. 1 (Autumn 1981). No. 2 (Spring 1982), No. 3/4 (Summer 1982), No. 5/6 (Winter 1982), No. 7/8 (Summer 1985—much delayed!). This tries to cover all aspects of the cube and some related puzzles. It is the only newsletter to cover the mathematical aspects of the cube. Much of the information in this chapter has appeared in more detail in the *Circular*. Copies can be obtained from me (David Sing-master).

Rubik founded a magazine called *Rubik's—Logic and Fantasy in Space*. There was a pilot issue in 1981 and eight issues appeared in 1982–83. It covered all aspects of spatial puzzles, but also covered general puzzles, games, art, etc. It appeared in five languages with quite a bit of colour. Back issues may be available.

Ideal Toy started a *Rubik's Cube Newsletter* in the United States in 1982, aimed at school level readers. I believe it had four issues.

At MIT, there was a Cube Lovers group and a mailbox for intercommunication. Several hundred pages of news and comments piled up in this from July 1980 to August 1982, which is when my copy ends. Unfortunately this is not readily available to other readers.

7.7   Prehistory and trials of the cube

In my Introduction to this book, I mentioned some earlier primitive cubes. William O. Gustafson did find a $2 \times 2 \times 2$ mechanism and patented it in 1963. It has an interior sphere with grooves. The 'impractical' 1970 patent was for a $3 \times 3 \times 3$ sphere with tongue-and-groove connections. Actually this has only 26 pieces—the interior is hollow. This was devised by Frank Fox, of Bletchley, England. However, UK patents must be regularly renewed and he let his expire in the mid-1970s.

The 'impossible' 1970 mechanism has turned out to be relevant. This was due to Larry D. Nichols of Arlington, Massachusetts, a chemist for Moleculon Research Corp. He describes a $2 \times 2 \times 2$ cubical puzzle with a magnetic mechanism. This is not like the idea which Rubik mentions in Chapter 1. (Incidentally, that mechanism was used in producing cubes in Japan.) Nichols has a bar magnet embedded in each inner face as shown in Figure 7.47. When one tries to turn a face, the like poles come together and force the pieces apart. Instead you must divide the cube in half, turn one half and then put the halves back together. This clearly violates the very first of Rubik's features, described in the Introduction. Consequently, Ideal Toy felt that Nichols' patent was not relevant.

However, Nichols' patent is quite general and allows for 'engagement by mechanical rather than magnetic means, as for example by using a pop-in snap linkage, or a tongue-in-groove arrangement' Also, Nichols' application preceded Fox's. Moleculon, who hold the

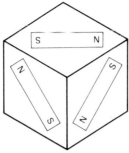

Figure 7.47

patent rights, filed suit against Ideal Toy for infringement of this patent in May 1982, asking 60 million dollars in damages. A major Washington law firm (Wilkes, Artis, Hedrick and Lane) took the case on a contingency basis — they get 25 per cent of the damages if they win and nothing if they lose. Nichols will get 20 per cent of what is left if they win.

The trial was in late January 1984 in Wilmington, Delaware, and lasted nine days. G. K. (Jerry) Slocum, the world's leading puzzle collector, was an expert witness for Moleculon for four days. In his opinion, the cubical shape with coloured faces was a major advance over Gustafson's 2×2×2 sphere. Gustafson himself has admitted that he had not realized the puzzle aspects of his device. Further, Slocum's study of the patent and puzzle literature showed that there was no other 'prior art' for this sort of puzzle. Although it seems obvious now to go from the 2×2×2 sphere to the 2×2×2 cube, neither Gustafson nor anyone else had done so before Nichols. Consequently, the court ruled, on 2 October 1984, that Nichols' patent was valid and that Ideal/CBS had infringed that patent.

However, the story is not yet over. CBS has appealed and the court has yet to decide on damages. There is no doubt that Nichols did not have an effective mechanism. His magnets make it too easy to disassemble the cube and hence cheat, or at least treat it as an assembly puzzle. This is borne out by the fact that Moleculon had tried to sell Nichols' cube to the toy trade in 1972 and it was rejected by all the companies they approached, including Ideal Toy.

Consequently, Rubik's mechanism represents a major improvement on Nichols' idea and must be considered as the principal part of the success of Rubik's Magic Cube. All parties are agreed that Rubik's work was completely independent. In view of this, I would predict that the court will award damages of only a small fraction of the amount claimed.

I personally feel that this rather complex and unresolved problem in no way detracts from Rubik's achievement. As I said in the Introduction to this book, the mechanism is the most wonderful aspect of the Magic Cube and I, and most other people, will continue to acknowledge this by continuing to call it Rubik's Magic Cube.

BIBLIOGRAPHY

Articles — Hungarian journals

Barabás, T. Hungarian horror (*Magyarország*, 19 April 1981)

Balázs, I. Portré kockával (*Magyar Hírek*, 4 April 1981)

Bedecs, É. A kocka kockázata (*Magyarország*, 25 November 1979)

Bognár, N. Kockakarrier, avagy az év játéka (*Magyar Ifjúság*, 28 November 1980)

Bóka, F. Egy játékszer világkarrierje (*Magyar Szó, Újvidek*, 15 October 1980)

Bossányi, K. Kocka—kockázat (*Népszabadság*, 15 October 1980)

Császár, I. A büvös kocka titkaiból (*Élet és Tudomány*, 16 May 1980)

Faludi, A. Mutatvány a büvös kockával (*Magyar Nemzet*, 30 December 1979)

Flesch, I. Büvös kocka világbajnokságot! (*Hétföi Hírek*, 8 February 1981)

Forró, P. Közbeszólas—kocka (*Magyar Hírlap*, 8 February 1981)

Fried, K. Büvös Kocka Klub (*Magyar Hírlap*, 9 March 1980)

Dr. Gantner, P. Büvös kocka—tanulságokkal (*Népiellenörzés*, October 1980)

Horváth, M. Kocka karrier (*Képes Újság*, 19 July 1980)

Huba, Z. Néhány szó a „Büvös kockáról" (*Számítástechnika*, May 1980)

Kaufmann, Z. A büvös kocka (*Fizikai Szemle*, 1980/5)

Keresztény, G. Beszélgetés Rubik Ernövel (*Észak.Magyarország*, 16 April 1980)

Kéri, G. A büvös kocka matematikája I.–II. rész (*Középiskolai Matematikai Lapok*, March and May 1980)

Kvassinger, K. Mindenki kockázik! (*Heti Világgazdaság*, 29 March 1980)

Maros, D. Büvös (*Népszabadság*, 13 July 1980)

Molnár, Zs. A büvös kocka—esettanulmány (*Ipari Forma*, January 1980)

Münz, A. Magyar Mágia (*Hétföi Hírek*, 27 December 1979)

Novobáczky, S. Mitöl büvös a kocka (*Ludas Matyi*, 23 April 1981)

Paládi, J. Kockakarrier (*Pesti Hírlap*, 21 September 1980)

Pálos, M. Játékbüvölök (büvös kocka—büvös golyó) (*Ország-Világ Karácsonyi Magazin*, December 1980)

Petö Gábor, P. A büvös kocka körül (1) ... 'elbüvölte a világot' (*Népszabadság*, 17 February 1981)

Rátonyi, R. Terefere—Rubik Ernövel (*Füles*, 17 April 1981)

Réti, P. Lépjen át a Rubikon! (*Heti Világgazdaság*, 7 March 1981)

Rubik, E. Nyílt tér (*Mozgó Világ*, 2 April 1977, 58 ... 59. o.)

Sáfrán, I. Kombinatorika (*Népszabadság*, 1 January 1981)

Somos, Á. A büvös (*Pajtás,* January 1980)
Somos, Á. Büvös Kocka (*Magyar Ifjúsag,* 29 July 1977)
Somos, Á. A játékos ember (*Magyar Ifjúság,* 30 September 1977)
Vadász, F. Öt büvös kockáról (*Népszabadság,* 17 October 1980)
Varga, T. A büvös kocka körül (2.) Matematikusszemmel (*Népszabadság,* 18 February 1981)
Varga, T. Még egyszer a büvös kockáról (*Élet és Tudomány,* 25 July 1980)
Büvös Magazin az Élet és Tudomány különkiadása September 1981

Non-Hungarian books
Bossert, Patrick, *You can do the cube.* Puffin Books, 1981
Endl, Kurt. *Rubik's cube (Strategie zur Lösung).* Würfel Verlag GmBH, 1980
Last, Bridget. *A simple approach to the magic cube.* Tarquin Publications, 1980
Nourse, James G. *The simple solution to Rubik's cube.* Bantam Books, 1981
Ogura, Seigi and Nakamura, Toshihiko. *You can arrange it in only two minutes* (in Japanese). K. K. Dynamic Sellers RT, 1981
Singmaster, David. *Notes on Rubik's 'Magic Cube';* published by the author, 1979 and by Enslow and Penguin, 1980
Taylor, Don. *Mastering Rubik's Cube.* Penguin Books, 1980
Shimauchi, Toichi. *A Rubik kocka mindentudója.* (in Japanese) Nihon Hyaron Sha. Tokyo, 1981.
Trajber, Josef. *Der Würfel 'Rubiks Cube'.* Falken Verlag, 1981
Warusfel, Ándre. *Réussir le Rubik's Cube.* Denoël, 1981

Non-Hungarian articles
Bauer, N. A race to conquer the Cube (*The Boston Globe,* 26 July 1981)
Blackman, T. Rubik's Cube city's latest rage (*The Gazette,* (Montreal) 4 December 1980)
Bronson, G. Turn for the Worse: 'Simple' Little Puzzle Drives Millions Mad (*The Wall Street Journal,* 5 March 1981)
Bureau, J. 'Pas si simple qu'hongrois' (*Liberation,* 3 May 1980)
Castel, J-M. Cube Infernal (*Le Meridional—le France,* 25 August 1980)
Chastain, S. Cubes that drive 'em plane crazy (*Philadelphia Sunday Inquirer,* 15 March 1981)
Chastain, S. Tortuous toy offers hours of frustration or fun for its fans (*The Kansas City Star,* 5 April 1981)
Constantinov, I. Vengerskii kybik (*Nauka i zizn'*)
Davenport, E. Erno Rubik's Maddening Cube (*International Herald Tribune,* 27 March 1981)
Desfayes, J. B. Le de mystificateur (*Le Nouvel Illustre,* 4 February 1981)
Ernst, S.. Würfelfieber greift auch auf die Schweiz über (*Tages-Anzeiger Zürich,* 19 July 1980)

Fink, M. Das Würfelwunder (*Trend,* No 3, March 1981)

Firon, A. Des Würfels dritte Seite (*Budapester Rundschau,* 16/1981)

Geai, J. P. Un cube diabolique part a la conquête du monde (*Forum* 29 April 1980)

Grant, J. Solving the Rubik's Cube shortage puzzle (*San Jose News,* (California) 4 March 1981)

Hadorn, W. Sucht um acht Ecken (*Schweizer Illustrierte,* 2 February 1981)

Halberstadt, E. Cube hongrois et théorie des groupes (*(Pour la Science,* No. 34, 1980)

Hofstadter, D. R. The Magic Cube's cubies are twiddled by cubists and solved by cubemeisters (*Scientific American,* March 1981)

Hughes, G. Mike is 20-second king of the cube (*Daily Mirror,* 12 August 1981)

Hutchinson, D. The many sides of Professor Cube (*Daily Mail,* 19 August 1981)

Jullien, P. Le cube hongrois (*Education et Informatique,* No 4, 1980)

Kidd, J. Frustration comes cubed (*Macleans* (Toronto), 6 April 1981)

Kirst, R. Das Gedulds spiel in der dritten Dimension (*Frankfurter Allgemeine Zeitung,* 13 June 1980)

Kousbroek, R. Het onzichtbaarheidscriterium (*Cultural Supplement NRC Handelsblad,* 20 February 1981)

Löwy, H. Ungarisches "Würfelfieber" nach österreich importier (*Arbeiter Zeitung,* 23 October 1979)

Marchionni, C. C. Cube craze multiplies in U.S. puzzle circles (*Lifestyles, USA,* 11 August 1980)

Meier, F. Rubik's Cube Club: to solve or solve not (*Stanford Daily,* 20 January 1981)

Miller, L. P. Rubik's What? (*News - Free Press,* 18 January 1981)

Meyer-Herlin, G. Zauberwürfel für lange Mussestunden (*Landsberger Tagblatt,* 10 May 1980)

Milloy, M. 43 252 003 274 489 856 000 That's how many potential questions Rubik's Cube poses. As for answers . . . (*Newsday,* 20 July 1981)

Mitchell, W. J. Hungarian Cube puzzle has cornered the market (*The Detroit Free Press,* 15 March 1981)

Mitchell, W. J. Hungarian's game creates cubic joy (*Chicago Tribune,* 2 April 1981)

Morris, S. Erno Rubik's Magic Cube (*OMNI,* 9 September 1980)

Montag, S. De kubus (*NRC Handelsblad,* 24 January 1981)

Muro, M. Trying to Solve Rubik's Cube (*The Boston Globe,* 25 July 1981)

Obermair, G. Fassen Sie das Ding nicht an! (*Süddeutsche Zeitung,* 12 July 1980)

McPeck, Ph. Rubik's Cube a puzzling metaphor (*The Sunday Pantagraph,* 12 April 1981)

Penix, L. Quixotic quest—Cubed (*The Cincinnati Post, Ohio,* 13 March 1981)

222   Bibliography

de Pracontal, M. Un cube fou, fou, fou! (*Science et vie*, No 753, June 1980)

de Pracontal, M. Cherchez l'algorithme de Dieu (*Science et vie*, No 764, May 1981)

Rijs, R. Wie temt de martelkubus? (*KIJK*, July 1981)

Rubinstein, S. Puzzle with a Twist (*San Francisco Chronicle*, 10 January 1981)

Savolainen, R. Rubik, kuutioja SM-Kilpailut (*Yhteislupa*, 19 May 1981)

Silcock, B. The craze for cubes (*The Sunday Times*, 15 March 1981)

Singmaster, D. Six-sided magic (*The Observer*, 17 June 1979)

Singmaster, D. The Magic Cube, (*Games & Puzzles*, No. 76, Spring 1980).

Singmaster, D. The Hungarian Magic Cube (*Mathematical Intelligencer* 2:1(1979).)

Thole, B. Würfelspiel mit verflixten Tücken (*Bosch-Zünder Zeitung*, 3 September 1980)

Tricot, J. Manifestations de groupe (*Jeux et Stratégie*, No 6, December 1980)

Tucker, G. The secret of Rubik's Cube (*The Light* (San Antonio), 25 February 1981)

Wade, J. Cube crazy! Brain teaser has put us all in a twist (*The Sun*, 27 May 1981)

Warshofsky, F. Rubik's Cube: Madness for Millions (*Readers Digest*, May 1981)

Werneck, T. Der Würfel des Teufels (*Frankfurter Rundschau*, 11 April 1980)

Additional books in English

[The following are the most extensive among the dozen of books which appeared in English since 1981—DS.]

Bandelow, Christoph. *Inside Rubik's Cube and beyond*. Birkhäuser, 1982

Frey, A. H., Jr. and Singmaster, David. *Handbook of Cubik Math*. Enslow, 1982.

INDEX

Names in the Bibliography are only included here if the name occurs elsewhere in the book.

223